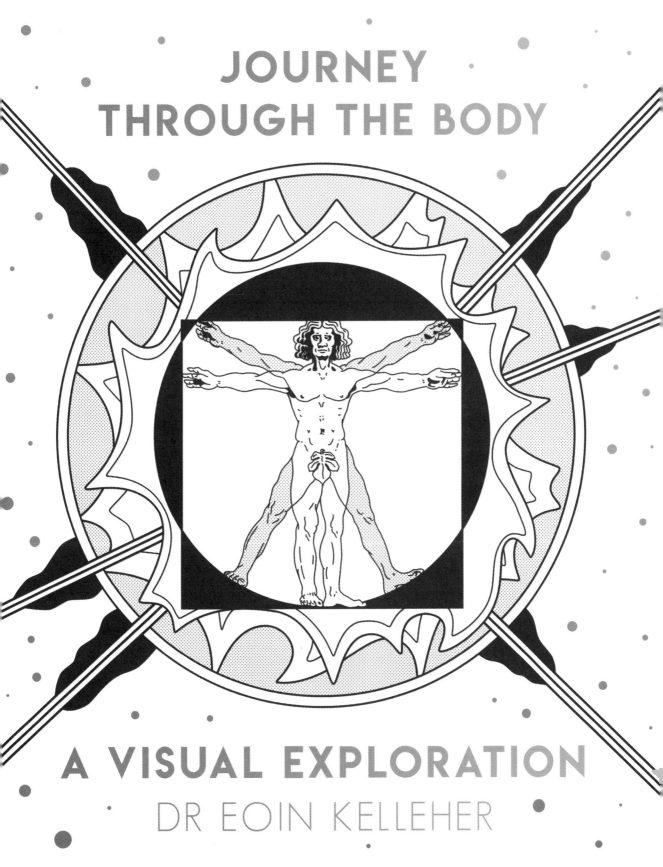

JOURNEY THROUGH THE BODY

A VISUAL EXPLORATION

DR EOIN KELLEHER

FOR EVERY BODY

MERCIER PRESS

Cork

www.mercierpress.ie

ISBN: 978 1 78117 708 2

A CIP record for this title is available from the British Library.

Printed and bound in the EU.

CONTENTS

INTRODUCTION

There is a world inside all of us that keeps itself hidden. While we may sometimes feel that we stumble or falter in our daily lives, our bodies are gracefully working with consistency and purpose. Behind the scenes, there are a whole variety of systems and processes to help us meet the challenges of the outside world.

When we stop and think about it, most of these systems do not require us to stop and think. We instinctively know, for instance, which way is up and which is down, even with our eyes closed. I would wager that if you put down this book momentarily and found a tiger facing you, you would not have to think to tell your heart to beat faster or your pupils to dilate. These things just happen automatically.

Our bodies are remarkably well designed for the world we inhabit. Our immune system is a stunning, complex creation that has evolved to protect us from the array of threats in the environment. Our bodies are adept at extracting large amounts of nutrients from the food we eat and converting them into the building blocks essential for life, even as the bones and muscles that scaffold us provide a robust and versatile means to navigate our world.

But the human figure also represents something more to us than the mere mechanics of survival. For millennia, humans have experimented with different ways of representing or depicting it. The human form is used to represent a myriad of deities. Parts of the body have been used to symbolise emotions, such as the heart for love, or the brain for intelligence. The subject of the human body has been a favourite of artists for many centuries, and this fascination has helped remarkably improve our knowledge of human anatomy.

With the following illustrations, I hope to shed light on some of the more interesting aspects of the body, from the intricacies of the immune system to our ability to interpret and adapt to the world around us. Artists or art movements have inspired many of these illustrations, but some are just idle creations of my imagination. I have not set out to write a comprehensive tome of physiology and anatomy, but merely to give you a novel glimpse into the inner workings of the ultimate work of art: the human body.

THE BRAIN

One of the greatest achievements of neuroscience over the past few decades has been identifying which areas of the brain are responsible for different functions.

The cerebrum is the largest and most developed portion of the brain. It can be divided anatomically into four lobes: frontal, parietal, temporal and occipital. Each lobe contains distinct areas responsible for different functions. The frontal lobe contains centres for speech, motor control and decision-making. The parietal lobe has the centres for sensation. The temporal lobe has the areas responsible for hearing, memory formation and spatial awareness. Visual perception is contained within the occipital lobe at the back of the brain.

Beneath the cerebrum lies the cerebellum, which is responsible for coordination and balance. Finally, connecting the rest of the brain to the spinal cord is the brainstem, the most primitive part of the organ. The brainstem is responsible for essential functions such as the control of breathing, regulating our heart rate and blood pressure, temperature control, sleep and other vital bodily functions.

The brain makes up about two per cent of a human's body weight and the cerebrum makes up eighty-five per cent of the brain's weight. However, overall brain size does not correlate with level of intelligence. For instance, the brain of a sperm whale is more than five times heavier than the human brain, but humans are still considered to be of higher intelligence.

DEEP CORTICAL STRUCTURES

Life on Earth is thought to have begun deep in the ocean near hydrothermal vents, and today some of the most bizarre living creatures reside in the deep sea. Similarly, while most of our conscious activity happens at the surface of the brain, in the cerebral cortex, a lot more is happening deep within the brain.

These areas are known as the deep cortical structures and include the cingulate cortex, amygdala, mammillary bodies, hypothalamus and thalamus. They are responsible for many tasks. For example, they act as a relay centre, managing the signals being sent out and received by the brain. They also coordinate the movements and signals being sent out to the rest of the body.

Santiago Ramón y Cajal was a Spanish pathologist and the father of modern neuroscience, who received the Nobel Prize in Medicine in 1906. He was the first to discover that the brain is made up of many individual nerve cells. In the course of his work he produced many detailed drawings of nerve cells that are now regarded as works of art in their own right.

AUTONOMIC NERVOUS SYSTEM

Many of the behind-the-scenes functions of the body are carried out by the autonomic nervous system, which is controlled by nerves in the brainstem and spinal cord. This system is made up of the sympathetic nervous system and the parasympathetic nervous system, i.e. the 'fight or flight' and 'rest and digest' systems.

The sympathetic nervous system can be stimulated by anything that might threaten survival. For example, if you were to put down this book and see a snake in front of you, your 'fight or flight' response would spring into action. Your pupils would dilate to improve your eyesight; you would breathe faster; your heart would beat more quickly; your mouth would go dry; you would go into a cold sweat; your hairs would stand up on your body; and blood would be diverted away from the gut towards muscles, thereby preparing you to fight or run.

When the parasympathetic nervous system is activated, so is our 'rest and digest' response. So if you happen to be sitting down with this book, you will be suitably relaxed for this response to kick in. Your pupils will constrict to focus on the text, your heart and breathing will slow down, and blood will be diverted to the gut to allow food to be digested and absorbed.

Roy Lichtenstein was an American artist known for being a leading figure in the Pop Art movement of the mid-twentieth century. By using industrial techniques to depict daily images taken from advertisements and comic strips, his work provoked a debate on the distinction between mass culture and fine art.

BODY CLOCKS

Circadian rhythms are physical, mental and behavioural changes that follow a daily cycle. (*Circa* is the Latin for 'around', *diem* for 'day'.) They respond primarily to light and darkness in the environment. Sleeping at night and being awake during the day is an example of a light-related circadian rhythm. Circadian rhythms are found in most living organisms, from animals and plants to tiny microbes.

In humans, our circadian rhythm is controlled by the master clock in a part of the brain called the suprachiasmatic nucleus, located in the hypothalamus, on the underside of the brain. However, almost every organ in the body contains its own independent clock as well.

The circadian rhythm is also influenced by outside factors, particularly daylight, as well as other factors known as zeitgebers (literally 'time giver' in German). Examples of these include mealtimes and room temperature. However, even if all these factors are removed, the human body is still subject to its own internal daily rhythm and there is nothing we can do to avoid it.

Surrealism was an art and literature movement that arose in Paris in the early twentieth century, with leading figures such as Salvador Dalí, René Magritte and Max Ernst. The Surrealists sought to liberate the mind from the constraint of rational thoughts and reason that dominated the Age of Enlightenment. The movement was influenced by the work of Sigmund Freud and placed a heavy focus on dreams and psychoanalysis.

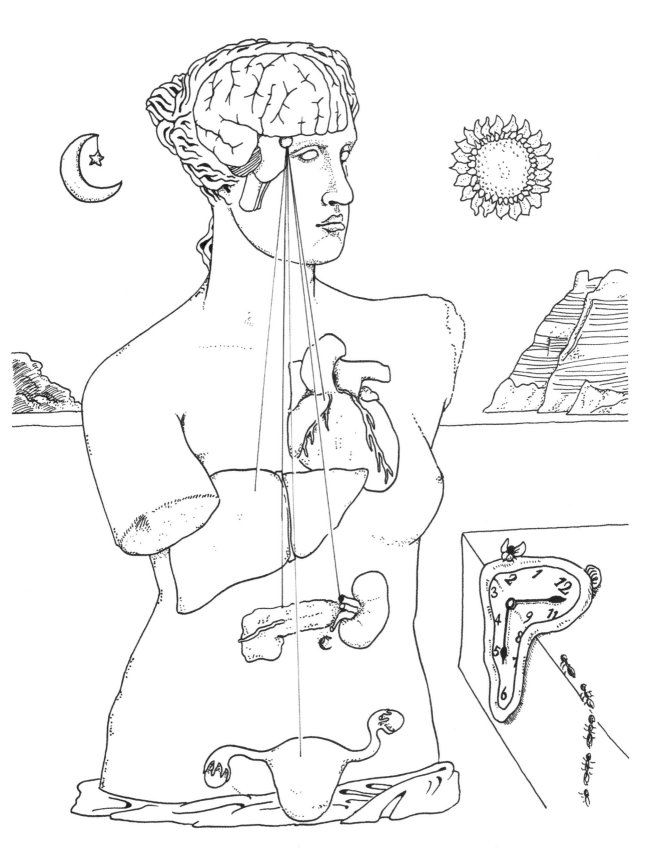

SLEEP

Every animal needs to sleep. Adult humans spend about a third of their lives asleep. Newborn babies sleep for around eighteen hours a day.

Sleep is one of the most important periods for the human body. It is essential for forming new memories, clearing waste products from the brain, maintaining a healthy immune system, and much more besides.

There are two stages of sleep: rapid eye movement (REM) and non-rapid eye movement (NREM). During REM sleep, the brain is very active, as this is when we dream. Fortunately, our muscles are normally paralysed during this stage, preventing us from acting out our dreams! It is during the NREM stage, however, that we are in the deepest level of sleep. It is at this stage that the brain consolidates and forms memories. However, because our muscles are not paralysed, it is also when sleepwalking can occur. If we do not get enough NREM sleep, we will not feel refreshed or rejuvenated in the morning.

The artist Salvador Dalí used to sleep holding a metal spoon and pot. When he fell out of REM sleep, the spoon would clang against the pot, making a noise that awakened him and allowed him to remember the dreams that inspired his Surrealist art.

CIRCLE OF WILLIS

The blood supply to the brain originates from four main arteries: two internal carotid arteries at the front of the neck and two vertebral arteries at the back of the neck. At the base of the skull, branches from these arteries form a ring (or a circulatory anastomosis) called the Circle of Willis. This arrangement is quite unlike any other pattern of blood vessel in the body and some would say that it looks quite alien!

If one of the arteries becomes blocked, blood from the other arteries can sometimes bypass the blockage by travelling along the Circle of Willis, thus preventing damage to the brain from a lack of blood supply. Thomas Willis was the seventeenth-century English physician who first described this arrangement.

The Circle of Willis is perhaps most well known because of its links to stroke. Strokes can occur when an artery gets blocked with plaque or a clot (called an ischaemic stroke), or when there is bleeding in or around the brain (called a haemorrhagic stroke). Small aneurysms (out-pouchings of blood vessels) are relatively common in the Circle of Willis, known as berry aneurysms. These can sometimes burst.

CORTICO-SPINAL TRACT

How does the brain transmit signals to the rest of the body? The answer lies in several bundles of nerve tissue called tracts that run from the brain through the spinal cord, from where individual nerves branch off. One of the most important of these is the cortico-spinal tract, which carries signals responsible for movement from the cerebral cortex (cortico-) to the body via the spinal cord (-spinal).

Interestingly, each side of the body is largely controlled by the opposite side of the brain. For example, your right hand is mostly controlled by the motor cortex in the left frontal lobe. The nerves travelling from one side of the brain cross over to the opposite side in the brainstem at a site called the decussation of the pyramids (decussation means 'to cross over').

Pneumatic tube systems are used to transport packages over short distances using a vacuum in much the same way that the tracts deliver nerve signals from the brain. Although early promise of use for transporting people did not amount to much, these systems are still present in many hospitals for transporting small packages, such as blood samples, to the laboratory.

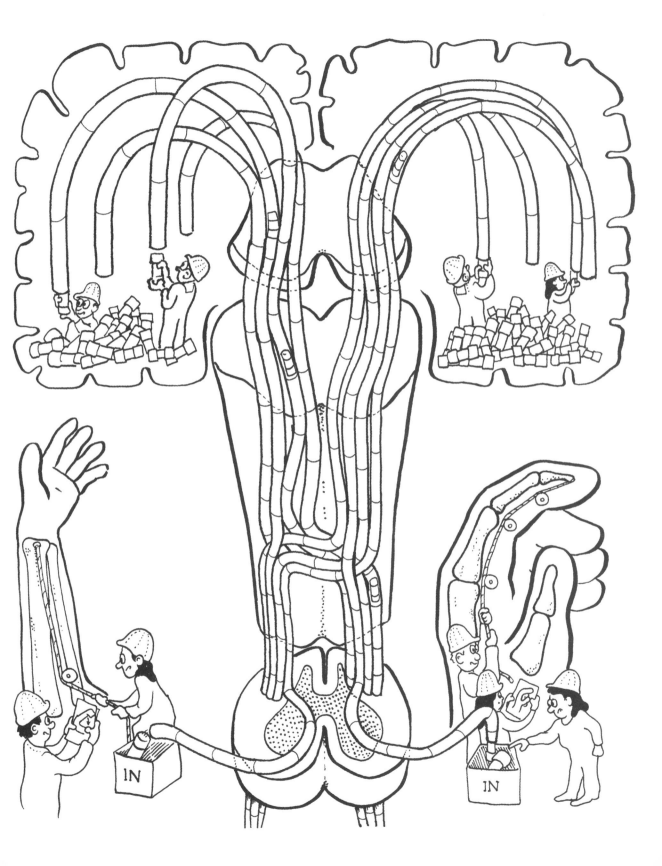

HOMUNCULUS

Not all areas of the body are afforded equal attention by the brain – some are more important than others. As a result, the body is often depicted in textbooks as a homunculus ('little person').

This little man is upside down, with oversized hands, feet and face (especially the lips and tongue). In contrast, the legs, arms and especially the trunk are shrunken. The differential sizes of areas within the homunculus represent the most and least sensitive areas of the body.

However, the homunculus is an over-simplification. In reality, the body is controlled by many overlapping areas of the brain; still, it remains an arresting image.

Some people who suffer from migraine with aura experience a weakness or tingling that spreads down one side of the body. This is caused by the spread of migraine activity along the area of the brain that the homunculus represents.

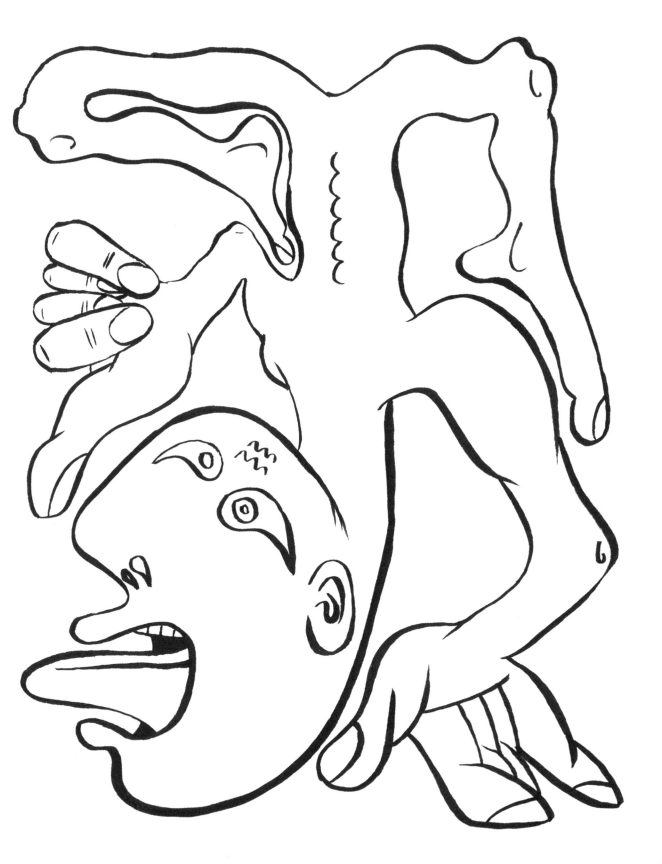

BRACHIAL PLEXUS

The arms are controlled by nerves that originate in the spinal cord in the neck. Five nerve roots join together to form a network of nerves – termed a plexus – that runs down behind the clavicle (collarbone) and into the axilla (armpit), where it then divides into the individual nerves that supply the arm.

The brachial plexus is vulnerable to injury during childbirth, particularly when one of the baby's shoulders gets caught above the mother's pubic bone. This is known as shoulder dystocia. If the brachial plexus is injured in this way it typically results in a pattern of weakness called Erb's palsy. It is also known as the 'waiter's tip' position. This is because the affected arm is rotated inwards with the hand facing upwards and backward, as if looking for a surreptitious tip.

DERMATOMES

A dermatome is an area of skin served by one spinal nerve. Each person has thirty dermatomes, which are arranged in segments that resemble slices through the body. This pattern is easily recognised on the torso, but on the arms and legs it is only evident when we are crouching on all fours – a legacy of our evolution from four-legged creatures.

Each spinal nerve is named according to the level that it exits the spinal canal. For example, the L2 spinal nerve exits below the second lumbar vertebra and receives sensory signals from the inner thigh. If an intervertebral disc slips and compresses the spinal nerve, it will be perceived as pain in that dermatome. Most slipped discs occur in the lumbar region of the lower back and often manifest as pain radiating down the leg. The pattern of the pain can help locate the source of the problem.

The varicella-zoster virus (best known as the cause of chickenpox) infects nerves, and after an acute infection can lie dormant in the cell bodies of nerve cells near the spinal cord. Many years later, the virus can reactivate and travel back down nerves to the skin, causing the characteristically painful rash called shingles. Because it is an infection of one spinal nerve, the rash is usually confined to one dermatome.

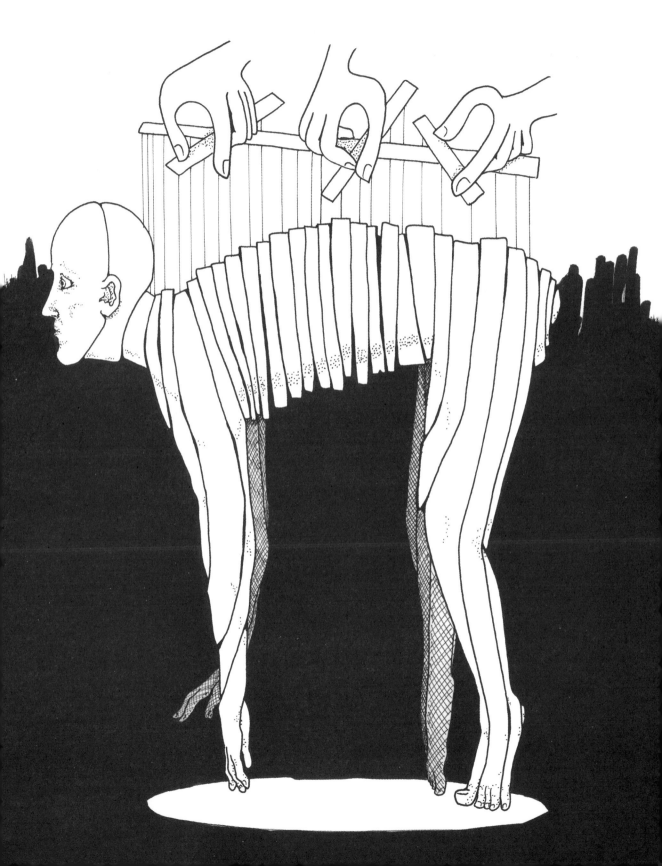

SALTATORY CONDUCTION

If nerve impulses had to travel in a continuous flow along nerves, like electricity in wires, they would be very slow (in fact, over ten times slower). Instead, nerve impulses jump along nerve fibres from point to point, which is a far more efficient method. The nerve impulses (called action potentials) jump between points called nodes of Ranvier in a process called saltatory conduction. (*Saltare* is the Latin for 'to hop'.) In between these nodes the nerves are insulated by sleeves of fatty tissue, known as myelin sheaths, which stop the nerve impulses from flowing through those points.

In people with multiple sclerosis the myelin sheaths that surround nerve fibres are damaged, resulting in loss of saltatory conduction. As a result, there will be slower transmission of nerve signals along nerves because the impulse now has to travel as a continuous flow along the entire length, much like the aforementioned electricity flowing through a wire. Put another way, it would be as if frogs had to swim the length of a pond rather than being able to hop across it on lily pads.

PAIN

The perception of pain is an essential bodily function. It alerts us to something wrong and to potential danger. Pain is a complex phenomenon. There are many different types of nerves that detect painful signals. These signals are relayed via the spinal cord to the brain, where they give rise to the conscious awareness of pain.

Because of the variety of signals, pain can be experienced in a multitude of ways: as sharp, stabbing or drilling; as electric shocks; as crampy or dull; as burning or cold; as tightness or twisting; and even as pressure.

However, pain need not always be a result of physical damage. The experience of pain over time itself can lead to a change in the way nerves develop, which can result in the sensation of pain even when the original injury has resolved, or where no injury existed in the first place. For example, many amputees experience pain long after the amputation has healed: both in the remainder of the limb and also in the absent limb itself (known as phantom limb pain).

Pain is also not just one-way traffic of signals from the rest of the body to the brain. Signals from the brain down to the rest of the body can alter how we perceive pain. This is why rubbing your toe after you stub it helps relieve the pain.

'The Wound Man' appeared in medieval medical texts. It usually portrayed an aggrieved individual bearing a variety of injuries and maladies. It acted as a table of contents, pointing readers to the relevant page that discussed the specific disease and its cure.

VISION

The sense of sight is a complex one. Light rays must enter the eye and hit the retina, where they are then turned into a nerve signal. This signal travels to the visual centre in the brain, which constructs the image that we see. However, what we see is heavily processed by the brain, which is one reason why we are susceptible to optical illusions.

These illusions occur because our brain is trying to interpret what we see and make sense of the world around us. Optical illusions simply trick our brains into seeing things that may or may not be real. For example, there is a blind spot on the retina of the eye that has no light receptors. We do not perceive a blind spot in everyday life, however, because our brain has adapted and fills in these gaps.

Pablo Picasso was probably the most influential artist of the twentieth century. Over his ninety-two years, he experimented and worked in many different styles. He is perhaps most famous for his Cubist works, such as 'Guernica'. By breaking down his subjects to different planes, Picasso sought to represent many different viewpoints of an object to better depict its three-dimensional form. It was a revolutionary break from the traditional use of a single fixed viewpoint that had existed since the Renaissance.

HEARING

The ear consists of three parts: the outer, middle and inner ear. Sound waves enter the outer ear and cause the eardrum (tympanic membrane) to vibrate. This in turn causes three tiny bones (called ossicles) in the middle ear to vibrate. These vibrations are transmitted to the inner ear (the spiral-shaped cochlea), where they are turned into a nerve impulse that travels to the hearing centre of the brain.

Concert halls are carefully designed and constructed to control the characteristics of sound within the room. The shape of the hall, and even the materials that line the walls, are important aids for the quality of sound that reaches the audience.

FLAVOUR

How we perceive the food we eat is determined by much more than just our taste buds. How food looks, smells and feels – its consistency and temperature – are all important. This requires the input of almost all our senses. In addition to taste, smell, vision and touch each play an essential role. These myriad signals travel to different parts of the cerebral cortex to determine our conscious interpretation of the food we eat.

The average person has about 2,000–5,000 taste buds on the tongue. Taste buds are sensory organs with very sensitive microscopic hairs called microvilli. These microvilli send information to the brain about how something tastes. It has been shown that brighter, more intensely coloured foods seem to taste better than less vibrant, duller-coloured foods, even when the flavour compounds are exactly the same.

SEMICIRCULAR CANALS AND BALANCE

One of our most important senses is balance. It requires input from two separate organs: semicircular canals and the otolith organs. Both of these lie in the inner ear.

The canals are responsible for detecting changes in rotation. There are three canals orientated at right angles to each other across three planes. Each canal recognises a rotation along different axes, which can be thought of as being similar to the axes that an aircraft rotates around. These are pitch (up/down), roll (side-to-side) and yaw (turning).

The semicircular canals work because of inertia (a resistance to change in a substance). Each canal is filled with a liquid called endolymph and contains a pocket of specialised cells called hair cells at its end. When your head turns, the fluid moves more slowly than the rest of the body, causing the hair cells to move in the opposite direction. This triggers nerve cells to send signals to the brain, telling the body of the change in direction.

Over time the fluid equilibrates with the rest of the body and the hair cells return to their original position and the signal fades. Thus, the semicircular canals only tell us about a change in position, not our actual position.

Because we have evolved to live on solid ground, aircraft pilots can be susceptible to spatial disorientation, which is where they are unable to determine their speed, orientation or altitude while in the air.

OTOLITH ORGANS AND BALANCE

The other parts of the body responsible for balance are the two otolith organs, also located in the inner ear. These are responsible for determining our body's change of velocity, or linear acceleration.

The two otolith organs are small sacs arranged at right angles to each other. They are lined with hair cells and contain many loose crystals (hence the name: *oto* means 'ear', while *lith* means 'stone'). Similar to the semicircular canals, they also function because of inertia.

As the body speeds up while moving in a straight line – for instance, while in a rocket taking off – the crystals move more slowly than the rest of the body because of their inertia. This acts on the hair cells and sends a nerve signal to the brain. This is how your body recognises it is moving in a straight line.

Balance is sometimes called the sixth sense. In fact, our bodies have many more than just five senses. The original five senses have their origin in the teachings of ancient philosophers, who ascribed senses to the elements. Aristotle believed there to be five senses, which he set out in his treatise *De Anima* ('On the Soul'). Sight was associated with water (because the eye contains water), hearing with air, smell with fire, and touch with earth. Taste was considered another form of touch.

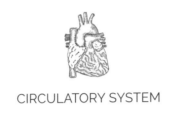

THE HEART

The heart is a pump that works all day, every day, for as long as you are alive. It is durable and efficient. Every minute, it pumps roughly five litres of blood to both the lungs and the rest of the body. The heart is divided into right and left sides, each containing two chambers – one atrium and one ventricle.

There are two phases of pumping. First, the atria contract and pump blood into the ventricles. Second, the ventricles contract and pump blood out of the heart. Blood from the right ventricle goes to the lungs. The left ventricle supplies the rest of the body.

The heart contracts in a twisting motion directing blood towards the aorta (carries blood to the body) and pulmonary artery (carries blood to the lungs).

The heart can respond very rapidly to how much blood the body needs. For example, if the veins return extra blood to the heart it causes its muscular walls to be stretched. This stretch alters the mechanics of the muscle cells, causing them to contract more strongly, thus pumping the extra blood out. These alterations are known as Starling's Law, after a scientist, Ernest Starling, who carried out a lot of research in the area.

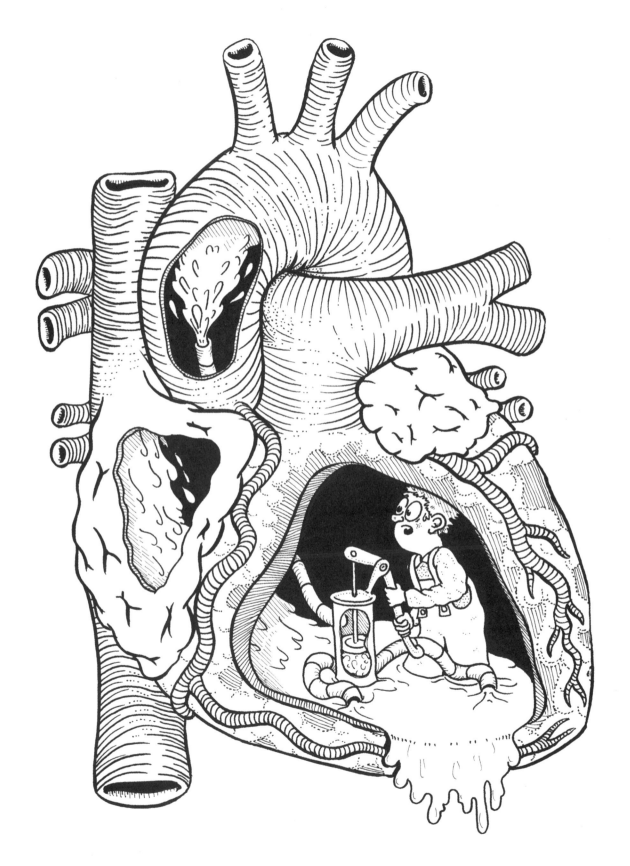

HEART VALVES

There are four heart valves, two in each side of the heart. If the heart did not have valves, blood would be sent backwards as well as forwards, clogging up the circulation. These valves billow out from the heart with the force of the blood flow, resembling parachutes in the process.

The mitral valve directs blood from the left atrium to the left ventricle, and the tricuspid valve directs blood from the right atrium to the right ventricle. The aortic valve directs blood from the left ventricle to the aorta and to the systemic circulation. The pulmonary valve directs blood from the right ventricle to the pulmonary artery and to the pulmonary circulation. The opening and closing of these valves generate the 'lub-dub' sound you hear when listening to your chest.

The mitral valve gets its name because it resembles the traditional Roman Catholic bishop's headgear, called a mitre.

CARDIAC CONDUCTION SYSTEM

Special types of muscle cells transmit electric impulses through the heart tissue, causing it to contract. As with a series of dominoes, the activation of one cell causes the next cell to be activated.

The electrical impulse begins in a group of special cells called the sinoatrial node, so-named because it lies on the wall of the right atrium beside the sinus venarum. The sinoatrial node acts as a pacemaker by activating at regular intervals that are determined by the amount of blood that the body needs at a particular time.

The impulse then travels through the atria and to the ventricles via the atrioventricular node, so-called because it lies between the atria and ventricles. The atrioventricular node slows down this electrical impulse, so that the atria have time to contract and fill the ventricles before the larger chambers are activated.

Atrial fibrillation is the most common heart rhythm disturbance. It occurs when chaotic electrical activity develops in the atria, taking over from the sinoatrial node. As a result, the atria no longer beat in an organised way and therefore pump blood less efficiently. The atrioventricular node will stop some of these very rapid impulses from travelling to the ventricles, but the ventricles will still beat irregularly and possibly rapidly,

CORONARY CIRCULATION

Not only does the heart continuously pump blood around the body, but it must also supply itself with the enormous amounts of oxygen needed to keep its muscle cells contracting. The heart is supplied with blood by the right and left coronary arteries. They originate from the base of the aorta and spread out over the surface of the heart.

Unlike most other organs, the heart consumes the majority of oxygen delivered to it. As a result, cardiac muscle is very vulnerable to any reduction in its blood supply from, for example, a blockage in the coronary arteries. Damage to the heart muscle caused by a reduction in the delivery of blood or oxygen is colloquially called a heart attack.

It is quite challenging to depict a three-dimensional structure, such as the heart, on a two-dimensional surface, such as a piece of paper. It is like peeling an orange and stretching the peel out on the table. Mapmakers face similar problems when making maps of the globe, resulting in distortions in the appearance of countries and continents.

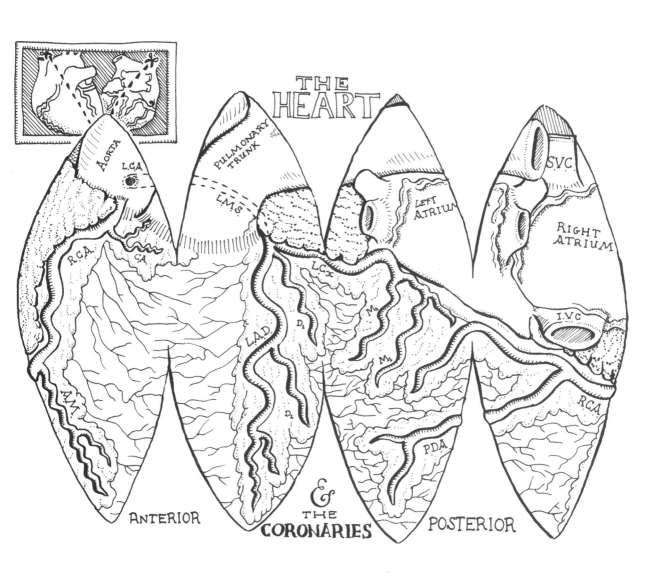

THE HEART

&
THE
CORONARIES

ANTERIOR · POSTERIOR

WILLIAM HARVEY

Before William Harvey, an English physician educated in Padua in the seventeenth century, most physicians followed the teachings of Galen. Galen was a physician-philosopher from Ancient Rome who set the pace for medicine in Europe and the Islamic world until the Renaissance. His followers believed that blood was created anew from food and then absorbed by tissues. They believed that the lungs were responsible for moving blood, and that the heart's function was the generation of heat.

However, Harvey's experiments demonstrated that blood was continuously pumped around the body by the heart. His findings were a revolution in medical science, but they were greeted with scepticism and it was decades before they were accepted. At the time, Harvey's findings were considered as startling and revolutionary as Copernicus' discovery that the Earth revolves around the sun.

Because the heart is the centre of the circulatory system, surgery on it can be quite tricky; unsurprisingly, it is difficult to operate on a beating heart that is full of blood! However, a machine can take over for the heart and lungs during surgery, taking blood directly from the large veins, oxygenating it and pumping it back to the arteries. This is called cardiopulmonary bypass and allows the heart to be stopped so it is easier to operate on.

THE CIRCULATORY SYSTEMS

The body has two parallel circulatory systems: the systemic circulation and the pulmonary circulation.

The systemic circulation originates from the left side of the heart and carries blood (with oxygen) to the rest of the body. It also returns oxygen-depleted blood back to the right side of the heart. Here, the pulmonary circulation begins and carries blood to the lungs, where it takes up oxygen before travelling again to the left side of the heart, starting the process all over again.

Sometimes both circulatory systems may be connected in an abnormal way. For example, a patent foramen ovale (PFO) is a hole in the heart between the right and left upper chambers (atria) that did not close the way it should have after birth. It is present in one-quarter of the population but does not usually cause any problems.

ARTERIAL SYSTEM

The systemic arterial system carries oxygenated blood (blood replete with oxygen) from the left ventricle of the heart to the rest of the body. The major artery of the body is the aorta. It leaves the left ventricle and arches over the heart, giving off major branches to the arms and head before travelling down the chest and into the abdomen. Before entering the pelvis, the aorta splits into two major arteries that supply the legs.

Arteries divide into smaller vessels called arterioles. Arterioles divide into microscopic capillaries. Oxygen and other nutrients are delivered to the body's tissues across the thin walls of capillaries.

Like pipes, arteries sometimes become blocked, for example by a clot or build-up of plaque. Sometimes the blockage can be removed, but if it cannot it may need to be bypassed. Arteries from another part of the body can be used to create a bypass around a blocked segment. For instance, arteries from the arm can be used to bypass blockages in the arteries supplying the heart.

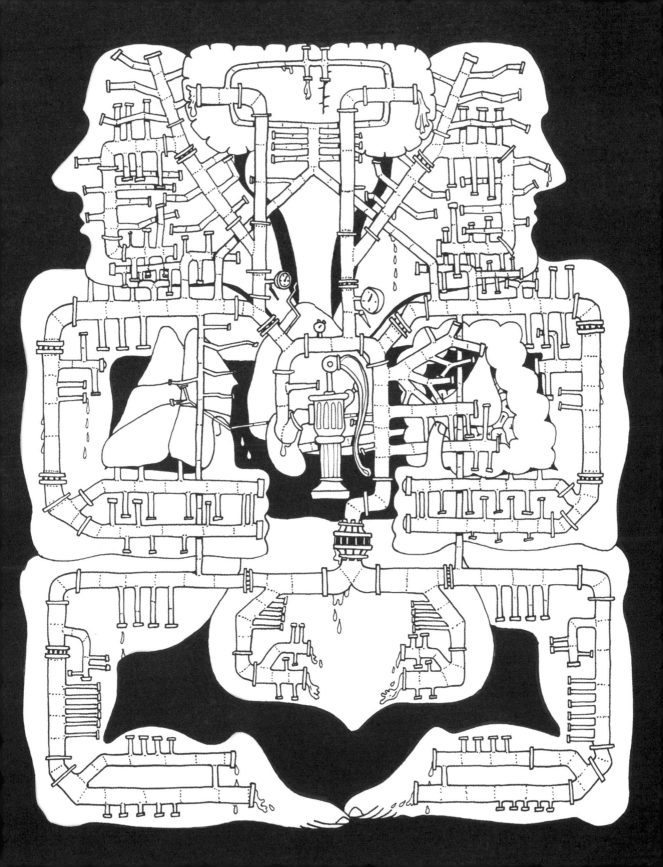

VENOUS SYSTEM

The venous system brings deoxygenated blood (blood lacking in oxygen) from the body back to the right atrium in the heart. From there it is pumped to the lungs to be refilled with oxygen.

Most blood passes through just one capillary bed before entering the venous system, although the body has a few exceptions known as portal venous systems. The most important of these is in the liver. Deoxygenated blood from the gut is carried to the liver by the portal vein, where it enters another capillary bed known as sinusoids. In this way, products of digestion are metabolised in the liver before being sent out to the rest of the body.

The veins are an important reservoir of blood. At any one time about two-thirds of blood is contained in the veins, compared to only a fifth in the arteries – the rest is in the heart and lungs. In times of stress, this blood can be mobilised by our sympathetic nervous system, or 'fight or flight' response. For example, if we were to sustain an injury and start bleeding, the body's sympathetic nervous system would cause the veins to contract, returning blood to the heart.

LAMINAR FLOW

Throughout most of the circulatory system, blood flows in smooth paths in concentric layers. This is known as laminar flow. It is the same pattern air follows when flowing over an aircraft wing. It is the most efficient way for a liquid to travel.

However, if there is a blockage in the path, the blood flow can become turbulent, or disordered. This generates sound as energy is lost, which can be detected. For example, if one of the heart valves is too narrow, blood flow will become turbulent and generate a noise known as a heart murmur.

When we breathe, the airflow to our lungs is also usually laminar. However, during an asthma attack the airways narrow and become swollen and clogged with mucus. This narrow and irregular pattern brings about a turbulent flow and increases the effort required to breathe.

In order to combat this during a severe asthma attack, a gas called heliox – a mixture of helium and oxygen – can be used. This is effective because a less dense gas, like helium, favours laminar flow, and so by inhaling it into your airways you improve the flow and make breathing easier.

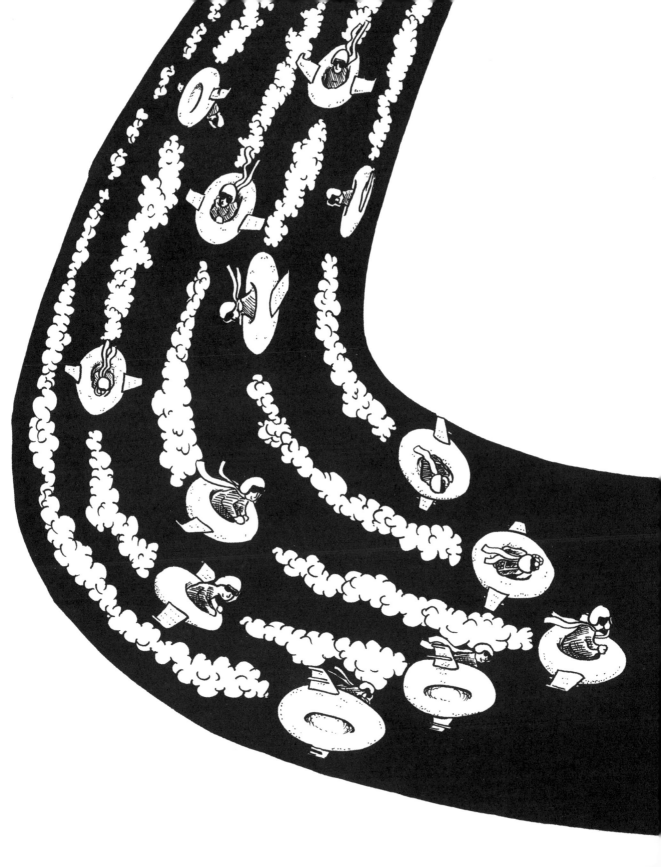

OXYGEN CARRIAGE

Red blood cells are dedicated to the job of carrying oxygen. As these hard workers develop from precursor cells, located in the bone marrow, they become more and more specialised, losing their cell nuclei until all that remains is a flexible sack full of haemoglobin. Haemoglobin is an iron-containing protein that enables red blood cells to carry plenty of oxygen. Flexible cell walls are needed by these oxygen-carrying cells in order to squeeze through the capillaries, some of which are narrower than the red blood cells themselves.

The heart pumps the red blood cells to the lungs, where they pick up oxygen, and then onwards to the other organs, where the oxygen is released. After around three months, the walls of the red blood cells start to become stiff. To prevent old red blood cells from simply breaking apart, they are ingested by white blood cells called macrophages in the spleen and liver. The iron from their haemoglobin is then sent back to the bone marrow for recycling.

Diego Rivera was a Mexican artist best known for his large murals depicting the lives of working-class people. He was married to another well-known Mexican artist, Frida Kahlo.

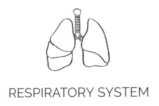

THORACIC ORGANS

The thorax is the area within the human chest bounded by the diaphragm below and surrounded by the ribcage.

Above, the thorax meets the neck in the thoracic inlet through which a number of important structures pass. The trachea, or windpipe, runs down the front of your neck and is covered with rings of cartilage, which gives it a bumpy feel. Overlying the trachea in the neck is the bow-tie-shaped thyroid, which is an important part of the endocrine system. On either side of the trachea run the blood vessels from the heart carrying blood to the brain.

On either side in the thorax itself are the lungs, each lined by pleura and largely covering the heart (although they are unhappily pulled away here). In between the lungs lies the mediastinum, a division within the thoracic cavity that contains the heart, thymus and great vessels. The great vessels include the aorta, the main artery that exits the heart and arches its way over and back before descending into the abdomen below.

The thymus is a fatty gland that sits on top of the heart. It is responsible for the production and development of T-lymphocytes, an important component of the immune system. Most of its activity happens in childhood, and it reaches its peak size during puberty. By the time you are an adult it begins to shrink and becomes little more than fatty tissue.

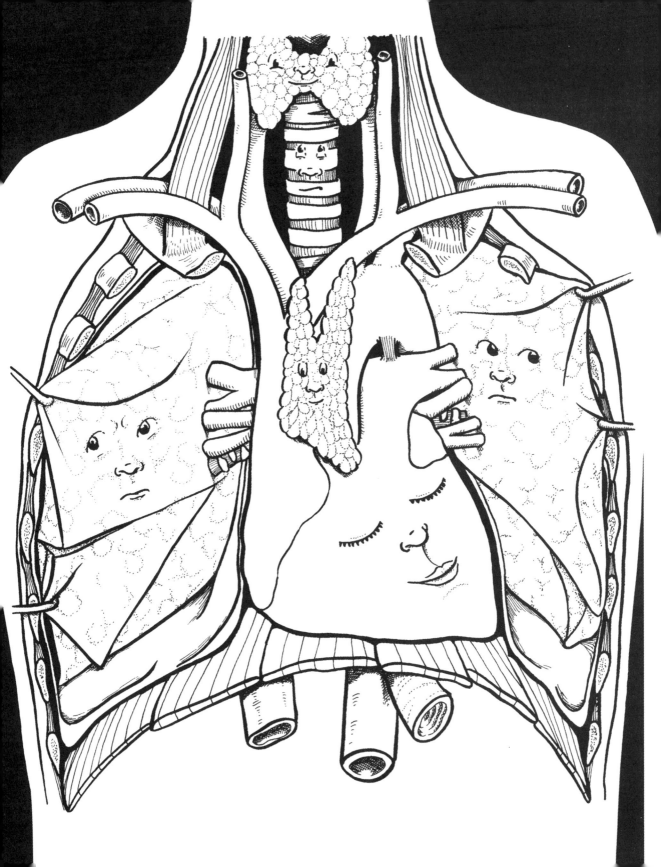

THE NOSE

The nose is more than just the organ of smell. Like a cave, from the entrance at the nostrils the nasal cavity extends deep into the face and connects with the throat. It is divided in two by a thin septum, and has three pairs of bones jutting into the cavity from either side. These are known as conchae because they resemble shells. Connecting to the nasal cavity are several other air-filled cavities called the paranasal sinuses. The sinuses make the head lighter, act as shock absorbers protecting the brain and, together with the nose, are important for speech and singing by giving resonance to the voice.

The nose and sinuses are lined with cells that produce large amounts of sticky mucus. The large surface area and warm, moist environment humidifies and warms the air we inhale. This prevents the airways from drying out. The mucus, together with nose hairs, also catches any large particles and stops them from reaching the lungs. If unwanted particles enter the nose, it triggers a sneeze reflex that quickly ejects them.

At the roof of the nasal cavity lie the nerve endings of the olfactory nerve, which is responsible for our sense of smell. There is only a thin plate of bone separating the nose from the brain here, called the cribriform plate.

While the nose stops most large particles from reaching the lungs, smaller ones evade its grasp. Nitrogen oxides (NOx) – a group of pollutants in the air released from motor vehicles, among other sources – are a particular nuisance.

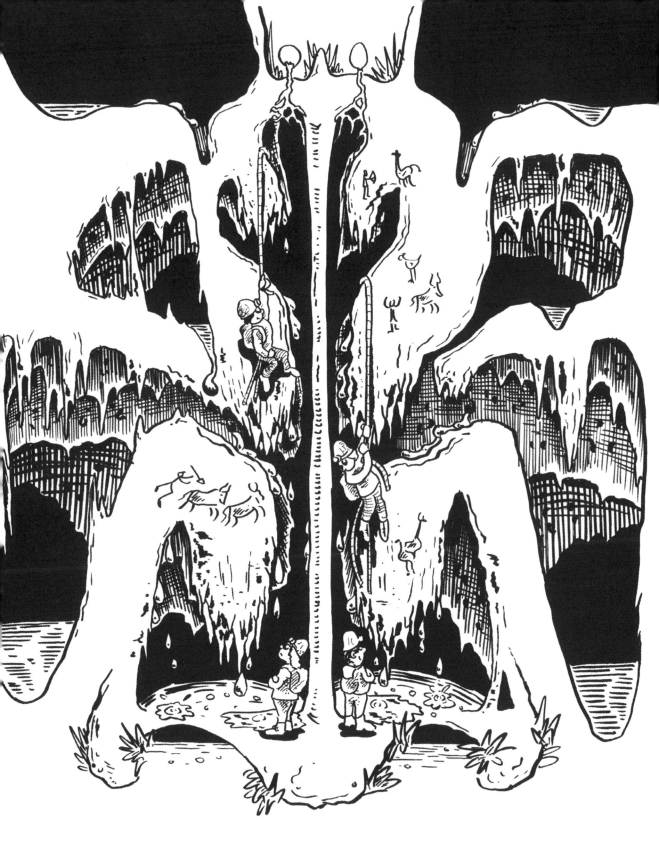

TRACHEO-BRONCHIAL TREE

When viewed upside-down, the lungs look like a tree. Starting with the trachea, the airways divide into progressively smaller branches called bronchi and bronchioles.

The airways end in a bloom of sac-like alveoli, where oxygen is absorbed into the blood and carbon dioxide is transferred from the blood to the air.

The right bronchus is straighter than the left. As a result, if you inadvertently inhale a foreign body it is more likely to end up in the right lung than the left lung.

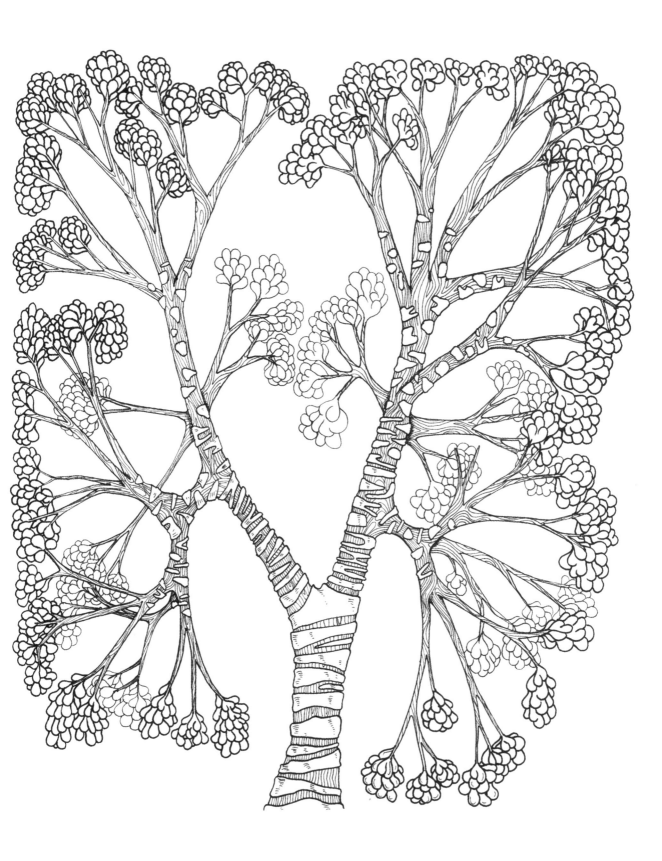

THE ALVEOLUS

The alveolus is where the action happens in the lungs. It is where oxygen contained in the air we inhale is exchanged for carbon dioxide dissolved in our blood.

Alveoli are found at the end of each bronchiole. They are roughly spherical and expand when filled with air, similar to balloons. However, if the alveoli were the same as balloons we would find breathing very strenuous. Inflating an empty balloon can take a bit of effort, but thankfully our bodies have developed to make breathing easier than that.

A substance called surfactant coats the inner surface of the alveoli and reduces the amount of surface tension. Surface tension is caused by the attraction of nearby water molecules to each other. It tends to collapse the alveoli. By reducing surface tension, surfactant makes alveoli easier to inflate as we inhale and stops them collapsing completely during exhalation.

Just as the Greek Titan Atlas held up the heavens, surfactant holds up the alveoli.

Babies born prematurely have not had a chance to develop surfactant and so they find breathing extremely difficult because of the work required to overcome the surface tension in the alveoli. However, the premature lungs can be treated with surfactant given directly down a breathing tube into the newborn's lungs.

GAS EXCHANGE

The body's tissues rely on oxygen (O_2) to enable them to efficiently release energy from carbon-based compounds such as glucose. Carbon dioxide (CO_2) is generated as a waste product of this reaction. Oxygen and carbon dioxide must travel to and from the various parts of the body in the blood, and there must be some means for the blood to acquire oxygen and expel carbon dioxide.

Each alveolus is like a dock for our body where this exchange happens. Blood vessels surround the alveoli, separated by only a very thin membrane, across which gas can pass down a concentration gradient. Thus, oxygen inhaled passes into the deoxygenated blood, and carbon dioxide from deoxygenated blood passes into the alveoli and into the breath that we then exhale.

Carbon monoxide poisoning occurs when too much carbon monoxide is inhaled, for example from cigarette smoke or indoor stoves. Carbon monoxide is harmful because it binds much more strongly to haemoglobin than oxygen. As a result, less oxygen is delivered to the organs, causing damage and a sensation that would feel similar to being smothered. People with carbon monoxide poisoning have a characteristic 'cherry-red' appearance.

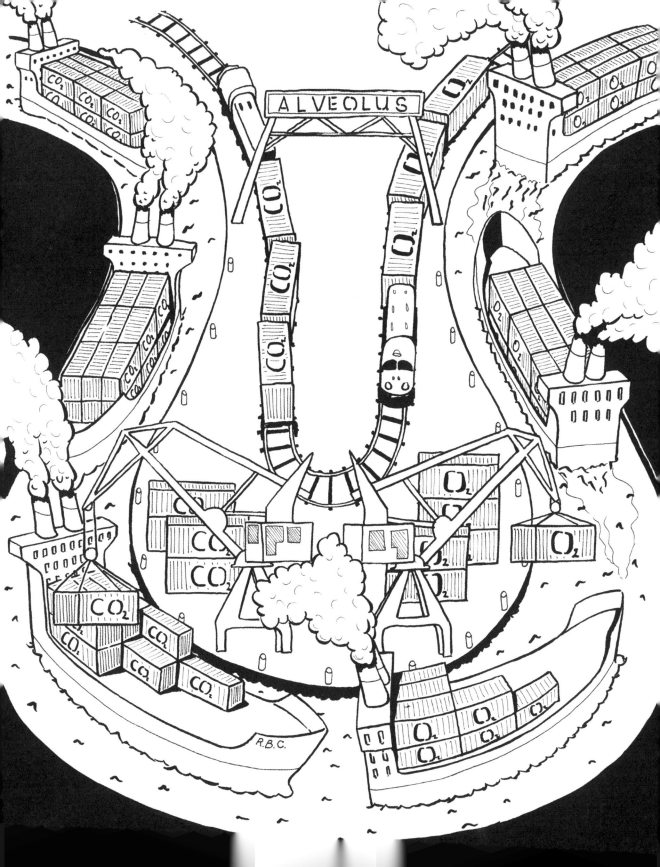

MUCOCILIARY ESCALATOR

It is not just oxygen that enters our airways when we take a breath. Dust, pollution, bacteria and other undesirables find their way in too. Thankfully, the body is able to expel these unwelcome visitors. First, they are trapped in sticky mucus that lines the airways. This mucus is then carried out of the airways into the nose and mouth by the movement of microscopic arms called cilia. Finally, this mucus is either swallowed or coughed or sneezed out. This whole process is called the mucociliary escalator.

When this system breaks down, the lungs quite quickly become infected and damaged. For example, in the genetic condition cystic fibrosis, the mucus in the airways is too thick to be cleared by the cilia. The resulting accumulation of mucus and bacteria leads to blocked airways and infections.

There is a rare genetic condition called Kartagener's syndrome where all the internal organs are located on the opposite side of the body (i.e. the heart is on the right side instead of the left, and so forth) and the cilia do not work properly.

COUGH REFLEX

Coughing is one of the body's important protective reflexes. It is essential that food, liquids and other substances do not go down the wrong way and end up in the lungs, where they could block airways or lead to inflammation and infection. The airways are extremely sensitive and the presence of anything that isn't supposed to be there will lead to a severe bout of coughing.

Whooping cough is a disease caused by the bacteria bordetella pertussis. It is also known as the '100-day cough' because the coughing can last for months. A person may cough with such vigour that they vomit, break ribs or even stop breathing. Although a vaccine for the infection was developed in the 1940s, the number of cases is on the rise, and it is a major cause of vaccine-preventable deaths worldwide, especially among young children.

THE ABDOMEN

There is a lot happening in the abdomen. With so many different organs all in one place, it can appear quite confusing. It is the site of much of the gastrointestinal system. In addition to important organs such as the stomach, liver and pancreas, there are almost eight metres of intestine crammed into the abdomen.

The abdomen is also culturally significant. Depending on where you live, a large midriff can be considered prestigious, because it signifies having abundant access to food, or shameful, as it represents a poor lifestyle. Similarly, there are many different attitudes to the display of the bare abdomen in public. For a long time in much of the Western world, public exposure of the bare abdomen, and the bellybutton in particular, was seen as indecent. In fact, the National Association of Broadcasters in the United States prohibited the display of women's navels (though not men's). This rule, although frequently flouted, stood until the 1980s.

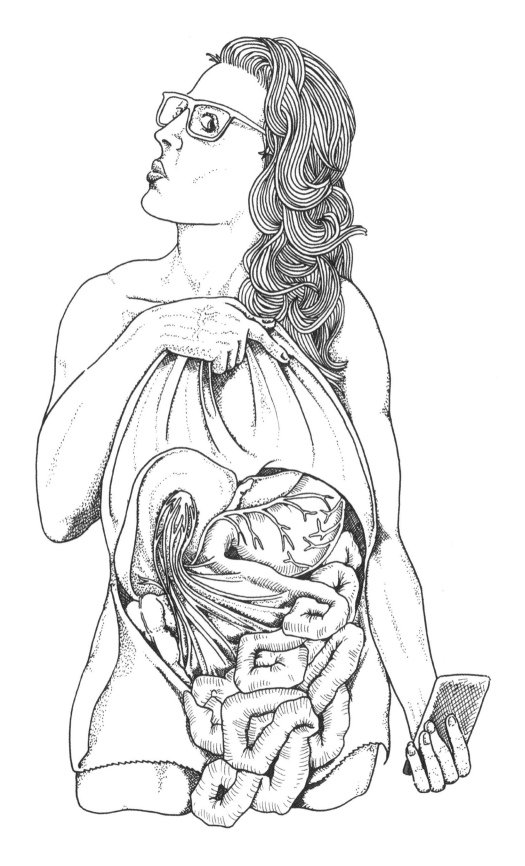

ABDOMINAL ORGANS

The abdomen is bounded by the diaphragm above and the pelvis below. The liver takes up the most room. It is nestled in the upper right corner of your abdomen (upper left side in the image, as you are facing another's abdomen) beneath the ribcage. The gallbladder is tucked in beneath it. The stomach (getting a bit of shut-eye) is formed from the oesophagus and lies on top of the pancreas. The pancreas in turn has three parts – a head, body and tail. The head is wrapped by the duodenum, the first part of the small intestine (duodenum means 'twelve fingers', a reference to its length). The tail of the pancreas reaches out to the spleen in the top left corner of the abdomen, almost tickling it.

Further back, beneath the liver on the right and spleen on the left, lie the two kidneys. The body's two major blood vessels, the aorta and inferior vena cava, enter the abdomen through the diaphragm and run down to just above the pelvis, where they both split in two.

The spleen is often regarded as a mysterious organ. However, it has many essential functions. It filters the blood and removes old or damaged blood cells. It also detects infections, especially bacteria, in the blood and stores lymphocytes which fight them. The size and location of the spleen can also be remembered by the old '1x3x5x7x9x11 rule': it is 1x3x5 inches, it weighs 7oz, and it lies behind the left ninth to eleventh ribs.

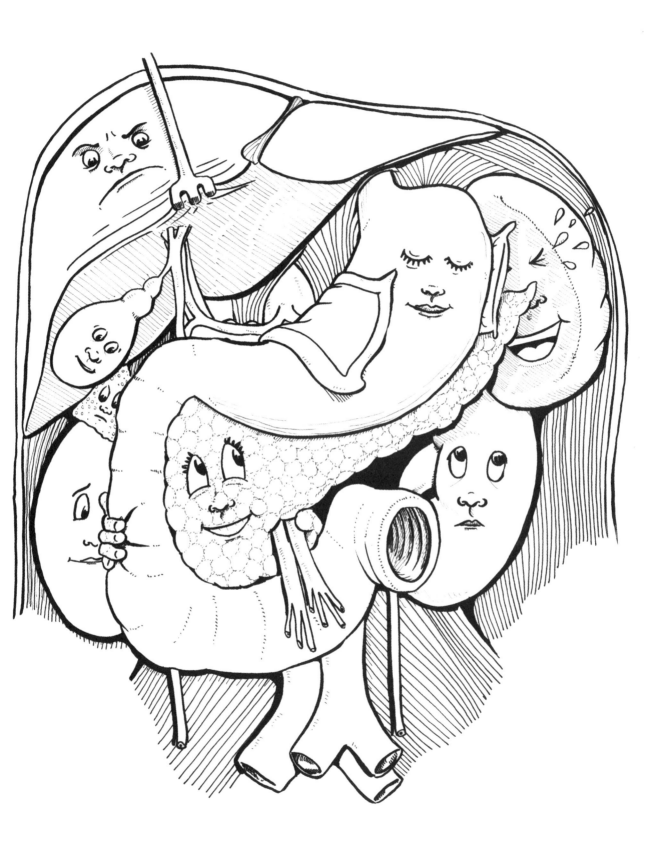

THE STOMACH

The initial stages of digestion help the gut extract as much nutrition as possible from the food we eat. Our teeth help separate our food into portions small enough to be swallowed. Enzymes in our saliva start to digest the food before it has even left our mouth. The act of swallowing passes the portion of food from the mouth through the oesophagus and into the stomach. There, gastric acid and enzymes break down the food further. The food is held in the stomach while it is churned by muscles located in the stomach wall. After several hours, it is released into the small intestine, where even more digestion takes place and the absorption of nutrients can begin.

Fritz Kahn was a German doctor and illustrator. He is best known for his imaginative mechanistic representations of the human body. One of his best-known works is 'Der Mensch als Industriepalast' ('Man as Industrial Palace'), which depicts the body as a sophisticated chemical plant.

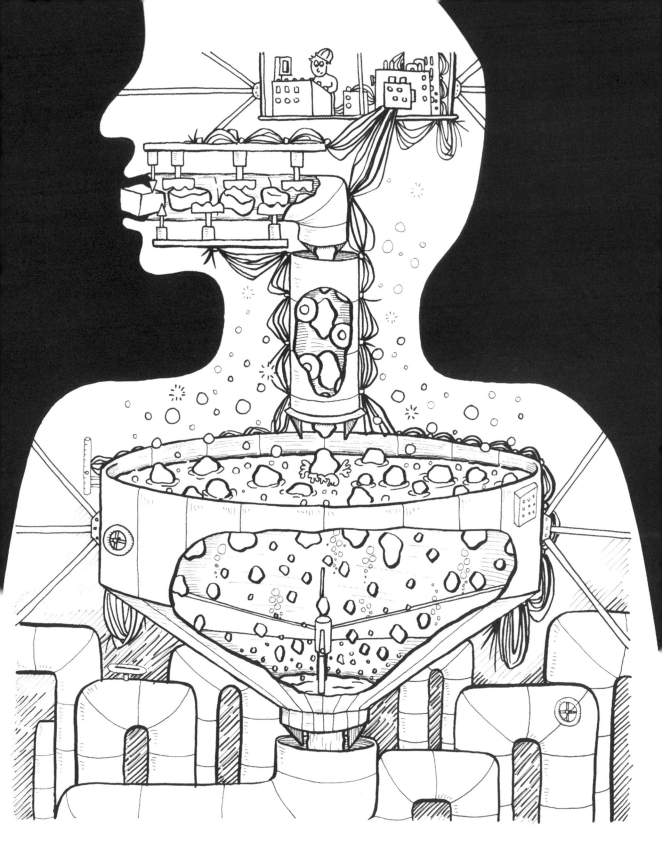

LIVER AND PORTAL CIRCULATION

The liver is the largest solid organ in the body and has a wide variety of vital functions. For example, it creates bile and other substances important for digestion. It makes proteins important for clotting and the immune system. It generates glucose and stores energy for later use in the body. It also protects us from harmful substances by processing many drugs and breaking down toxins.

The liver has a unique blood supply known as the hepatic portal system. The veins carrying blood from the stomach and intestines merge to form the portal vein, which enters the liver. Thus, all blood from the gut must first pass through the liver before reaching the rest of the body. This acts like a checkpoint for substances entering the rest of the body from the gut. The products of digestion are processed by the liver, where they are either released into the systemic circulation, stored for later use, or metabolised.

Any drugs that are ingested reach the liver via the portal system, where they may be processed. Some of these may be rendered inactive by the liver. However, if the same drug is given directly into the systemic circulation (for example, by injection, or even by inhaling), they can exert their effect. For instance, the general anaesthetic drug propofol is extremely potent if given into a vein, causing unconsciousness, but if it is swallowed it is inactivated by the liver and will have no effect.

OMENTUM

The omentum is an apron of fatty tissue that hangs from the stomach, covering the intestines. It plays an important role in protecting our gut. It has clusters of white blood cells known as milky spots dispersed throughout its surface. These act as a filter for harmful substances, such as pathogens, and can alert the immune system to potential threats.

The omentum was christened the 'abdominal policeman' by British surgeon James Rutherford Morison in 1906. He noted its propensity to adhere to damaged structures, such as perforated ulcers. Surgeons have taken advantage of the omentum to support repairs of intra-abdominal structures such as the intestines.

GUT MICROBIOME

Not all bacteria are bad. In fact, the vast majority of bacteria are beneficial. An enormous number of bacteria reside in our gastrointestinal system, especially in the large intestine. Collectively they are known as the gut microbiome.

In exchange for a warm place to stay and a plentiful supply of food, these bacteria carry out a number of helpful tasks. They help break down difficult-to-digest foods, such as fibre. They are also essential for producing building blocks for the body, such as proteins. Perhaps most importantly, the so-called 'good' bacteria help keep the 'bad' bacteria from causing us trouble.

You are more bacteria than human. Our bodies contain over 100 trillion bacteria. We have ten times more bacteria than human cells and they play a significant role in our health and longevity. Everyone has a unique gut microbiome.

ENTERIC NERVOUS SYSTEM

The brain and spinal cord are not the only organs with a lot of nervous tissue. Your gut has a brain of its own! The gut contains many nerve cells collectively referred to as the enteric nervous system. It is linked to, and can influence, the central nervous system, but can act entirely independently.

Its most important function is co-ordinating the movement of the gut while it digests food. However, it also appears to have a role in immunity and mood.

There are in fact more neurons in the gastrointestinal tract than the entire spinal cord. Neurons in the enteric nervous system produce chemicals called neurotransmitters, most of which are identical to the ones found in the brain. For example, more than ninety per cent of the body's serotonin, a chemical best known for promoting a sense of well-being, lies in the gut, not the brain. In addition, about half of the dopamine in the body also resides in the gut. Among its many functions, dopamine is linked to the brain's reward and pleasure centres.

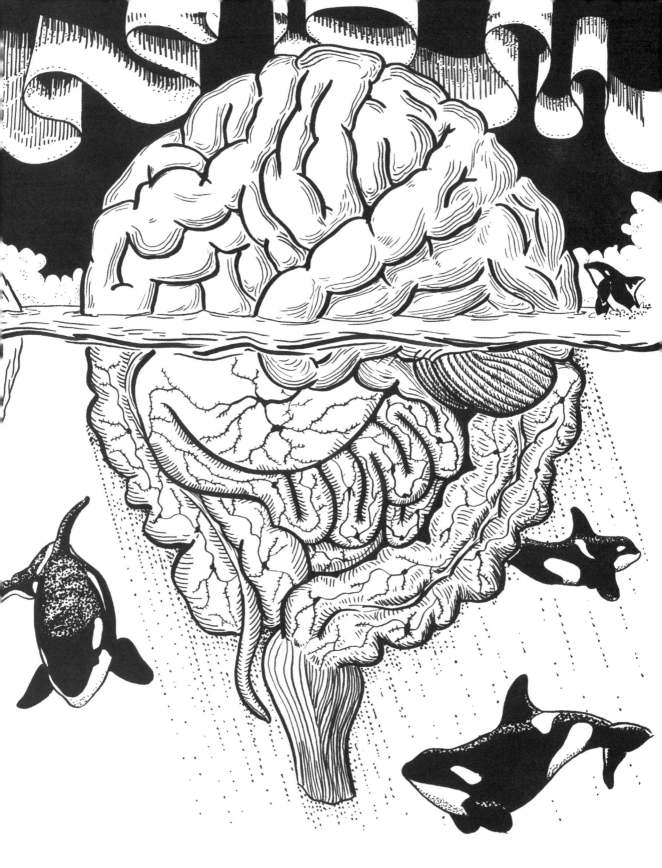

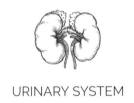

KIDNEY AND BLOOD PRESSURE

Our kidneys are very important for regulating our blood pressure. This role is essential, as the body must maintain an adequate pressure in the arterial system in order to perfuse all the organs and to allow filtration to occur (similar to the process of making espresso, where water must pass at high pressure through the ground coffee). The kidneys influence blood pressure by sensing the pressure at 'stretch' receptors near the glomerulus (more on this later) and then releasing hormones that increase blood pressure if it is too low.

High blood pressure (also known as hypertension) is a major cause of illness and death in the modern world; it often has no symptoms until it is too late. It can be caused by, but also itself cause, kidney disease. Worldwide, roughly one-fifth of deaths in men, a quarter of deaths in women, two-thirds of strokes and one half of heart disease are attributable to high blood pressure.

NEPHRON

The kidney is made up of hundreds of thousands of individual units called nephrons (from the Greek for 'kidney'). These nephrons serve to clear the blood of waste and ensure that the body maintains an adequate balance of water and salts. To enable the kidneys to perform these important functions, they receive a large proportion of the heart's output – one litre of blood per minute. If all the water contained in that blood were allowed to pass into the urine, we would need to drink an awful lot to keep up!

Glomerular filtration is the first step in making urine. Each nephron in your kidneys has a microscopic filter, called a glomerulus, that is constantly filtering your blood. During this filtration process, waste and extra fluid pass into the kidney tubules and become urine. Eventually, the urine drains from the kidneys into the bladder through larger tubes called ureters.

If your kidneys fail or stop working, this can lead to an accumulation of water and salt in the body, resulting in swelling. Sometimes people may require dialysis to replace the normal kidney function and filter blood using an artificial membrane that removes waste and excess water.

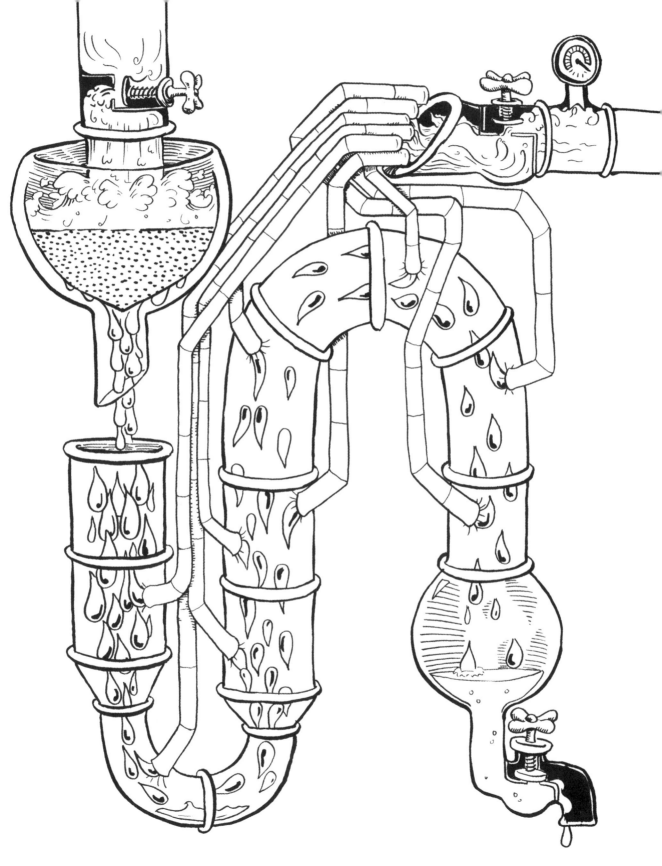

GLOMERULUS

The glomerulus is a tiny ball-shaped structure and is one of the most important parts of the kidney. It must allow waste products to be filtered out of the blood but prevent larger particles such as blood cells and important proteins from being filtered.

It achieves this selective filtration through a sophisticated three-layer struc- ture, with pores small enough to allow smaller particles to pass through. It also has a slight negative electrical charge, which repels negatively charged proteins in the blood and prevents them from being filtered.

If the glomerulus is damaged, the larger proteins will be able to pass through into the urine. Although this will not initially cause any symptoms, protein can be detected in the urine using a simple dipstick test and it is one of the first signs of kidney disease.

FLUID COMPARTMENTS

Osmolarity is the concentration of particles called solutes in a liquid solution. For example, salt in water.

The body is divided into different compartments that are separated by barriers that only allow certain particles through, usually determined by their size and electrical charge. They can be thought of as fields, separated by hedgerows with a fence. One field contains cows and sheep, while another contains no animals at first. Because sheep can jump over the fence, but cows cannot, the sheep will distribute themselves evenly between the fields but the cows will not. This is known as a concentration gradient, whereby solutes (or, in this case, sheep) move from an area of higher concentration to an area of lower concentration.

The kidneys depend on these concentration gradients because of their role in filtering the blood and maintaining the correct salt balance for the body.

If the kidneys stop working, another method must be found to filter the blood – this is known as dialysis. Dialysis does not have to be carried out with machines, however. A technique called peritoneal dialysis involves filling the lining of the abdomen (called the peritoneum) with a few litres of fluid each day. Solutes in the blood cross the peritoneal membrane down the concentration gradient into this fluid, which is then removed. This can help replace the filtering role of the kidneys by facilitating removal of excess fluid and waste products.

THE PELVIC CAVITY

The pelvic cavity lies below, and is continuous with, the abdomen. It is quite a crowded space. At the front lies the bladder, which is mostly behind the pubic bone. However, when full, it balloons upwards. The two ureters travel down from the kidneys in the abdomen to enter the cavity and deliver urine to the bladder. In females, the uterus lies partly behind and partly on top of the bladder, with the ovaries on either side. It expands in pregnancy and compresses everything else in the pelvis. At the back of the pelvis, snaking forlornly down from the abdomen, is the sigmoid colon and rectum.

The major blood vessels of the body also course through the pelvis. The abdominal aorta splits into two iliac arteries, and the inferior vena cava divides into two iliac veins. These vessels give off branches to the pelvic organs before travelling beneath the inguinal ligament to enter the thighs.

The ovaries and testes develop from the same structure, the gonadal ridge, in the abdomen. Ultimately, both the testes and ovaries descend from their original location, bringing their blood and nerve supplies with them.

SPERM CELL

Sperm are the male reproductive cells. Their function is to reach the ovum (or egg) and deliver the male's genetic material. A sperm cell is made up of a number of parts.

The head contains the nucleus with twenty-three chromosomes. At the top of the head is the acrosome, which is important in leading the sperm cell to find the ovum and also contains enzymes necessary to enter the ovum. Beneath the head is the neck, which contains the centriole. The centriole is important for development of the embryo.

Beneath the neck is the mid-piece, which has spirals of mitochondria (the cell's power plant) around a central core. These mitochondria produce the energy needed for the long journey to reach the ovum.

The tail, known as the flagellum, is the longest part of the cell. This produces the wave-like movement to propel the sperm cell to its goal.

Unlike a rocket ship, sperm cells do not operate entirely independently once they have taken off! Progesterone, one of the female reproductive hormones, stimulates sperm cells and causes them to move more rapidly.

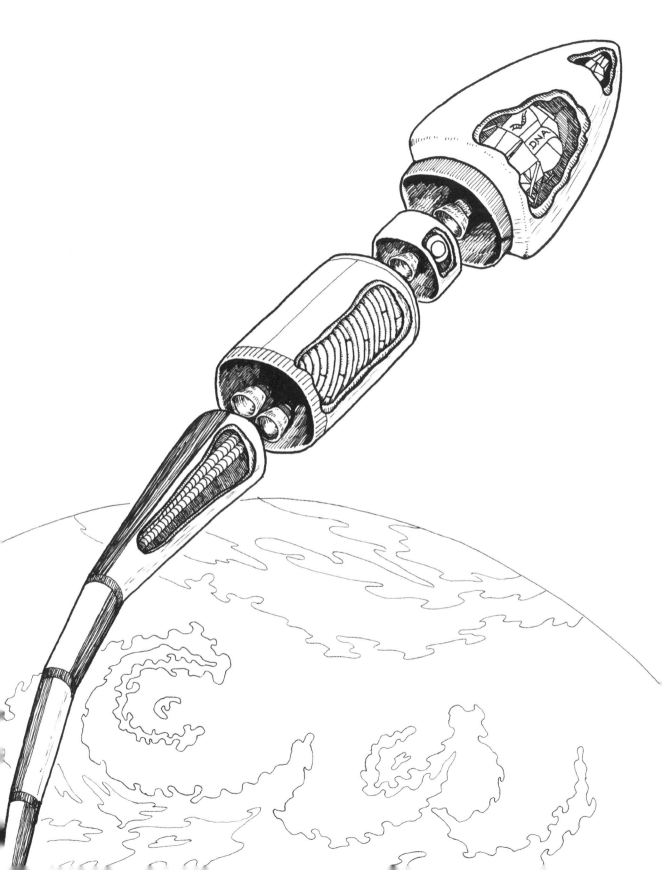

GUBERNACULUM

The testes develop in the abdomen, near the kidneys. However, as might be obvious to you, they end up far from there.

An embryological structure known as the gubernaculum (from the Latin for 'rudder') is responsible for this sequence of events. It is a ligament that, Moses-like, guides the testes from the warm abdomen to the relatively cool promised land of the scrotum! Along the way, they take their blood supply, lymphatic drainage and nerve supply with them. This is why the vessels supplying the testes originate from near the kidneys rather than in the pelvis.

Occasionally a testis can get lost along the way and fail to reach the scrotum by birth. This is called cryptorchidism, or undescended testes, and affects about one per cent of males. Undescended testes are more prone to problems such as testicular torsion, an exquisitely painful condition where the testis twists on itself and requires surgical fixation.

OVULATION

The ovaries contain millions of eggs, or oocytes, at birth. These eggs are contained in small pockets called follicles. The release of eggs from the ovaries is called ovulation. This occurs during the menstrual cycle, which has two phases: the follicular and luteal phases.

In the follicular phase of the cycle, these follicles are stimulated. The follicles grow until they rupture and the egg within them is released from the ovary into the fallopian tube. This marks the start of the luteal phase of the cycle. The remnants of the ruptured, and now egg-less, follicle then turns in on itself and becomes the corpus luteum, which produces oestrogen and progesterone. These hormones will support the development of the embryo if fertilisation occurs. If it does not, the cycle will start over.

Pomegranates are a symbol of fertility in many cultures. In Ancient Greek mythology, Hades, god of the underworld, kidnapped Persephone, the daughter of Demeter, the goddess of fertility, and took her to the underworld, where he tricked her into eating six pomegranate seeds. It was the rule that anyone who ate food in the underworld had to spend eternity there. Persephone's father, Zeus, commanded Hades to release her. However, since Persephone had consumed the six pomegranate seeds, it was agreed she would only be allowed to live on Earth for half of the year and would spend the other half in the underworld.

FERTILISATION

Fertilisation occurs when a sperm travelling in one direction in the fallopian tubes meets an egg (or ovum) travelling in the other direction and they fuse, forming a zygote.

The acrosome of the sperm cell contains enzymes essential to penetrate the tough outer membrane of the ovum called the zona pellucida. Once a sperm has fused with an ovum, another reaction takes place, resulting in the membrane of the ovum becoming hard and impenetrable to other sperm cells. This is to prevent more than one sperm fertilising an egg, though this does occasionally happen.

Conjoined twins are a rare type of identical twins. Identical twins form when a single fertilised egg splits into two separate embryos. When the egg fails to split fully, the result is a pair of conjoined twins. They used to be known as 'Siamese twins' because of a famous pair of conjoined brothers, Eng and Chang Bunker, from Siam (now known as Thailand).

ENDOCRINE ORCHESTRA

The endocrine system is formed from a diverse collection of glands that secrete hormones into the blood. Hormones are special substances that regulate and co-ordinate a wide array of bodily functions. These include growth and development, reproduction, sleep, mood, appetite, meta-bolism and much else besides.

Although it is very small, the pituitary gland is regarded as the conductor of the entire system. It is a small, peanut-shaped gland located on the undersurface of the brain, attached to the hypothalamus by a stalk. It secretes a host of hormones that regulate the other endocrine glands throughout the body.

The pituitary gland gets its name because it was thought by historical anatomists such as Galen and Vesalius to be where nasal mucus was produced. *Pituita* means 'slime' or 'phlegm' in Latin.

SECOND MESSENGER SYSTEMS

Hormones do not exert their effects directly. They travel in the bloodstream looking for target cells that have the receptor for the hormone on the surface. The hormone binds to the receptor and sets off a cascade of reactions in the cell through second messenger systems. In this instance, the hormone is the first messenger and other substances within the cell act as second messengers that amplify the response to the hormone.

In the illustration, the sun is the first messenger to herald the onset of day. Then the second messenger system begins as the light shines through a magnifying glass, which heats up the windowsill. This causes the frog to leap onto the weighing scales, tipping them. This pulls a string that closes the blades of the scissors. The watering can then tips forward, emptying its contents onto a watermill that turns a wheel, promptly smacking the sleeping man on the face, awakening him. Similar convoluted processes are at play inside all our cells.

Rube Goldberg was an editorial and satirical cartoonist in the United States in the early twentieth century. He was famous for designing absurd machines that executed simple tasks in a complicated manner; these contraptions simply became known as Rube Goldberg Machines.

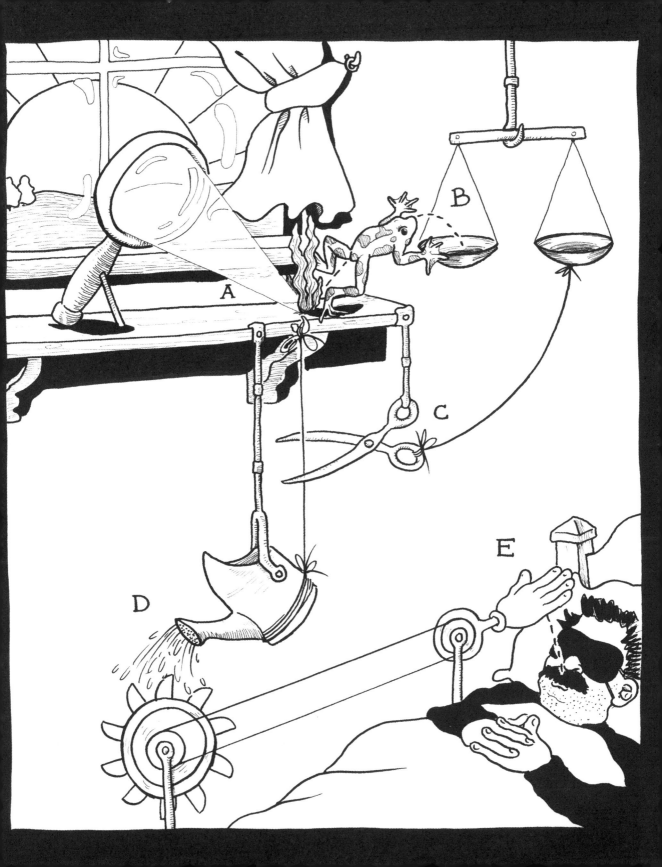

AMPLIFICATION OF HORMONE EFFECTS

Having a series of different hormones in a system allows for amplification, whereby a small amount of hormone can initiate a large response.

For example, thyroid-stimulating hormone (TSH) is released from the pituitary gland and acts on the thyroid gland to cause the release of many more molecules of thyroid hormone (known as T3 and T4). Thyroid hormone then moves through the bloodstream to exert its effects on the rest of the body, whether that be for regulating growth or metabolism.

The thyroid gland requires iodine to make thyroid hormone. If there is not enough iodine in the diet, a person may develop hypothyroidism, which can cause health problems. They may develop an enlarged thyroid gland, called a goitre, as it is stimulated to grow due to a lack of thyroid hormone. As a result, in many countries, iodine is added to food such as table salt, in an effort to prevent such issues.

THE IMMUNE SYSTEM

Our body is protected from harmful invaders called pathogens by the immune system. This is made up of a complex network of cells and organs that collectively seek out and destroy these pathogens. There are two main types of immune response: innate and adaptive immunity.

The innate immune system is the first line of defence. It is made up of the physical barriers facing the outside world, such as the skin and mucous membranes of the gastrointestinal and respiratory tracts. If these physical barriers are breached, for example with a cut or splinter, the white blood cells that make up a key part of the innate immune system are attracted to the area. This results in inflammation (and will be explained in more detail later).

The adaptive immune system is so-called because it is the body's means of adapting to threats over a lifetime. Specialised white blood cells, called T- and B-lymphocytes have the ability to learn and remember pathogens. If the same pathogens enter the body at a later time, the adaptive immune response swings into action and quickly dispatches them.

Military metaphors abound in disease. We battle illness, engage in a war on cancer, and undertake 'cohort' studies (from the Roman military unit *cohors*). They gained prominence with germ theory in the nineteenth century. Prior to this, disease was seen as an imbalance of bodily humours, something within us to be rectified. With germ theory, disease was now seen as something external to us, caused by outside invaders that had to be defeated.

BLOOD CELLS

All blood cells descend from the same ancestor – stem cells in the bone marrow – much like differing branches of the same family sharing a common ancestor.

The first major division of blood cells is into lymphoid cells and myeloid cells. Lymphoid cells develop into a type of white blood cell called lymphocytes. These are categorised as T-cells, B-cells or Natural Killer (NK) cells. These are important in the adaptive immune system.

The other blood cells develop from myeloid cells. These include red blood cells and other white blood cells, such as macrophages and platelets – though platelets are not, strictly speaking, cells. They are actually fragments of a larger cell, called a megakaryocyte.

Stem cell transplants are used to treat some diseases of blood cells, such as leukaemia, which is a cancer of the white blood cells. After powerful chemotherapy, and occasionally radiotherapy, stem cells are given to the patient in order to take up residence in the bone marrow and help make healthy blood cells again.

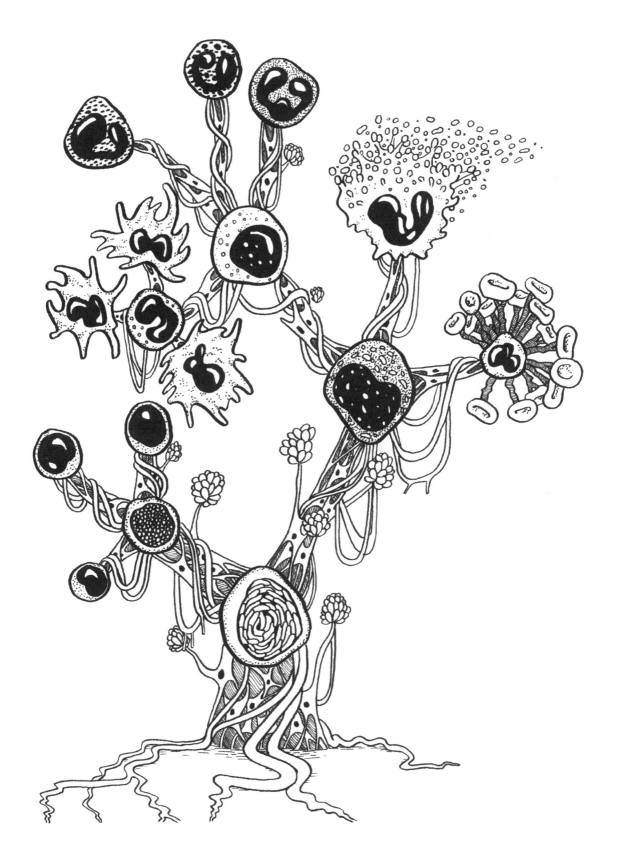

INFLAMMATION

Inflammation is the body's initial response to injury, and is part of the innate immune system. It is characterised by the cardinal features: heat, redness, pain and swelling. They are often described in medical textbooks by the Latin words *calor*, *rubor*, *dolor* and *tumor*.

When an injury occurs, signals released from damaged tissue cause the flow of blood to the area to increase. The extra blood brings an army of white blood cells that, in turn, produce the classic signs of inflammation.

Hieronymus Bosch was a painter from 's-Hertogenbosch in the Netherlands. 'The Garden of Earthly Delights' is one of his best-known works. It depicts Heaven and Hell as a chaotic landscape crawling with fantastical and surreal creatures, just as one might imagine the site of an invasion of the body and the ensuing inflammation.

MACROPHAGE

Macrophages are a type of white blood cell found in almost every organ. They patrol their part of the body looking for microbes such as bacteria, and other harmful elements such as cancer cells. Once found, pathogens are engulfed and destroyed. This process is called phagocytosis, from the Greek words for 'to eat' and 'cell'.

Macrophages then pass parts of the digested pathogens to other white blood cells called T-cells, so that the offending agent can be specifically targeted by the adaptive immune system.

Harry Clarke was an Irish illustrator and stained-glass artist who lived in Dublin at the turn of the twentieth century. He was diagnosed with tuberculosis (TB), a common illness at that time. Mycobacterium tuberculosis, the cause of the disease, infects macrophages. (As with other pathogens, macrophages ingest the bacteria that cause TB. However, macrophages are unable to digest TB and so the bacteria multiply inside them.) Clarke journeyed to a sanatorium in Switzerland for treatment; however, he died from the illness on his journey back to Ireland in 1931.

ANTIGEN-PRESENTING CELLS

The innate immune system is the first responder and indiscriminately targets any perceived threats. However, as previously mentioned, the body also has a more targeted, adaptive immune system. This system learns from its innate counterpart how to recognise threats.

This occurs through the innate immune system presenting broken-down parts of threats they have encountered to the cells of the adaptive immune system. These broken-down parts are called antigens. The adaptive immune system commits the antigens to memory, allowing it to respond faster if it sees them again.

William Blake was a poet and artist who lived in London at the end of the eighteenth century. His masterpiece is probably the illustrations for *The Divine Comedy* by Dante Alighieri, itself one of the great works of literature. Blake was regarded as an eccentric in his own time because of his focus on the mystical and supernatural, rather than the reason and rationality of the Age of Enlightenment. After his death, his reputation grew, and he is credited with inspiring the Romantics and Surrealists who came after him.

ANTIBODIES

Antibodies are a key component of the adaptive immune response. They are Y-shaped proteins produced by a B-lymphocyte. Each B-lymphocyte can produce one type of antibody against only one specific antigen. When that antigen is detected in the body, the B-lymphocytes for that antigen start to multiply and produce antibodies.

The two arms of the 'Y' protein act like hands and grab hold of the antigen and mark it as a target for phagocytosis (ingestion) by other cells in the immune system. This also serves as a trigger to activate the complement system (detailed overleaf).

This is the basis for many vaccines. An antigen is presented to the body (such as a piece of a virus, for example), giving the adaptive immune system an opportunity to recognise it and make antibodies against it. If the person then becomes infected by the virus, the body can mount a much faster and stronger immune response.

Antibodies are generated to protect the body against foreign substances, such as viruses and bacteria, but sometimes they can have a negative impact – for example, if they cause a rejection of a transplanted organ. This may occur if the recipient has pre-existing or develops new antibodies against the donor organ. The more differences there are between the donor and the recipient in terms of the kind of proteins (known as HLA, or human leukocyte antigen) that they have on the surface of their cells, the greater the risk of organ rejection.

COMPLEMENT SYSTEM

The complement system is part of the innate immune response. As the name suggests, it complements the function of antibodies and white blood cells, helping them recognise and rid the body of pathogens. The system is made up of a collection of proteins in the blood, rather than cells.

The system is constantly active at a low level, but the presence of an antigen causes a cascade that dramatically increases its activity. One effect of this is that complement proteins form something called a membrane attack complex, which punches a hole in the wall of a pathogen cell, destroying it.

Edvard Munch was a Norwegian artist who is best known for 'The Scream'. Munch described being out for a walk at sunset on a fjord overlooking Oslo when he experienced an 'infinite scream passing through nature'. The face he depicts has become one of the most iconic expressions in art, representing the anxiety and turmoil of the modern age.

ACTIVATION OF B-CELLS

Lymphocytes are the key components of the adaptive immune system. A T-lymphocyte that has been activated by an antigen-presenting cell from the innate immune system can then go on to activate B-lymphocytes. The activated B-lymphocyte multiplies and produces antibodies specific to the antigen it has encountered. Some of the B-lymphocytes will also produce memory cells, which grant the body long-lasting immunity by recognising the pathogen as foreign and mounting a response against it sooner when next exposed to it.

Michelangelo was commissioned by Pope Julius II to paint the ceiling of the Sistine Chapel. It contains over 300 figures across 500 square metres of painting. The best-known image is probably the 'Creation of Adam', depicting God's right arm outstretched to give the spark of life to Adam in much the same way that a T-lymphocyte would activate a B-lymphocyte.

COAGULATION CASCADE

The body does not contain a large volume of blood: only about 70ml for every kilogram of body weight. Therefore, it sensibly has a sophisticated mechanism to stop it from losing too much.

An important part of this safety mechanism is the clotting or coagulation cascade: a series of reactions involving proteins called factors. When a blood vessel is damaged, it triggers the cascade, resulting in a clot and attracting platelets to the area. Thus, the hole in the vessel is plugged.

A lack of any one of these clotting factors causes a condition called haemophilia, where there is a propensity to bleed easily. A rare genetic form of this condition, Haemophilia B, passed from Britain's Queen Victoria through various European royal families in the nineteenth and twentieth centuries. Haemophilia B is also known as Christmas disease, after Stephen Christmas, the first person in whom the deficient clotting factor was identified. The article describing the condition was also published in the Christmas edition of the *British Medical Journal* in 1952. Stephen Christmas tragically died at a young age from AIDS, contracted from contaminated blood products used to treat the disease. Similar afflictions affected many haemophilia sufferers around the world.

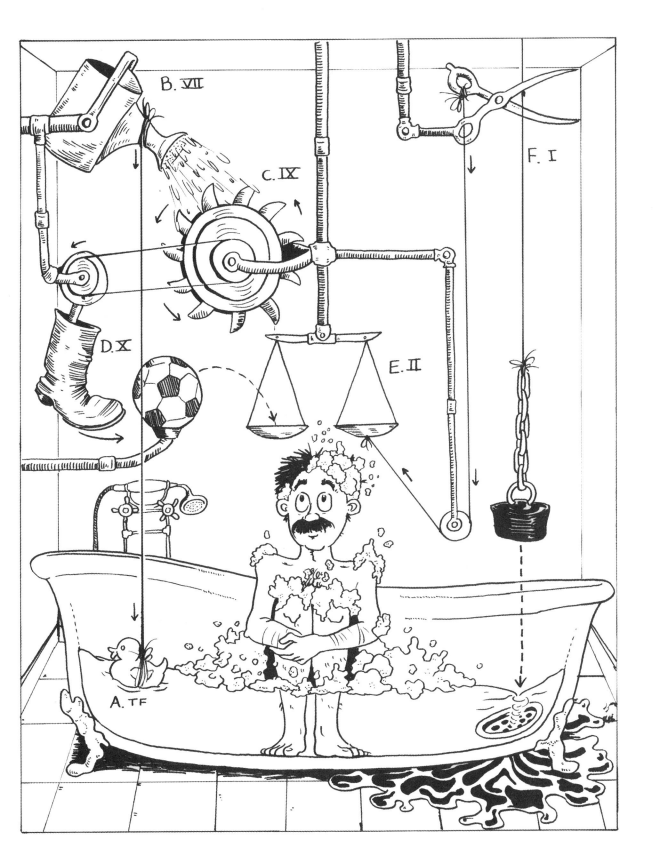

PLATELET AGGREGATION

There are three main cellular components in the blood: red blood cells, white blood cells and platelets. Platelets are essential because of their role in stopping bleeding.

Most of the time, platelets travel in the blood minding their own business. However, if the wall of a blood vessel is damaged, and the clotting cascade is activated, nearby platelets become 'sticky' and adhere to the vessel wall. They then roll along until finding the area of damage and join together or 'aggregate' to form a plug to stop the bleeding.

When your platelet count is low, you can experience easy or excessive bruising. This superficial bleeding into the skin appears as a rash of pinpoint-sized reddish-purple spots (petechiae), usually on the lower legs.

SKIN LAYERS

The skin is made up of two layers: the epidermis and dermis. Beneath these lie subcutaneous tissue and muscle.

The epidermis is the part of the skin you see. The outermost part of the epidermis is made up of dead cells. New skin cells are made at the bottom of the epidermis and move to the top layer, where they slough off and are replaced once again. Skin cells in the epidermis also make melanin, which gives skin its colour.

The dermis lies beneath the epidermis. It contains the roots of hairs, which are also attached to small muscles that contract to cause goosebumps. Sweat glands are bundles of curls in the dermis that produce liquid, which helps keep the body cool by evaporation. Nerve endings in the dermis give us our sense of touch and body position. Blood vessels not only bring oxygen and nutrients to the skin, they are also important in regulating body temperature.

Beneath the skin lies a layer of fat that attaches the skin to the muscles and bones.

Because of the weightless environment, astronauts lose the thick layer of toughened skin on the soles of their feet through disuse. After a long stint in space their feet resemble those of a newborn baby.

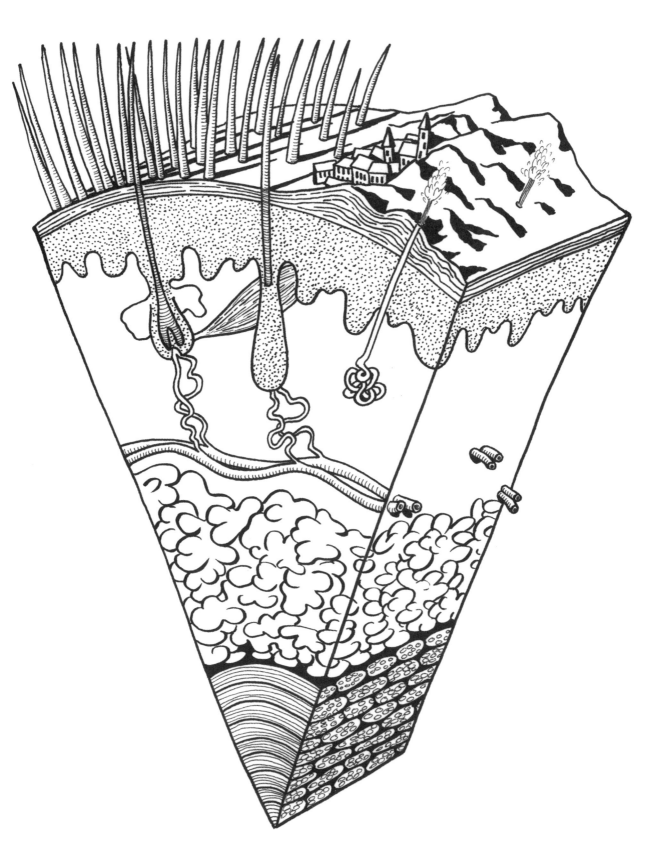

MUSCLE CONTRACTION

Muscles are comprised of many muscle cells arranged in bundles of fibres. Within each muscle cell there are many long bands called filaments. There are two types of filament: thin and thick. The thin filaments are spiral-shaped. The thick filaments are straight, with attachments called myosin heads.

Muscles contract when a nerve impulse causes calcium to be released from its storage place inside the cells of that muscle. The sudden increase in calcium causes each myosin head to attach to an adjacent thin filament and pull it along. This is similar to how a blade pulls a boat along while rowing, with the calcium acting as the coxwain shouting 'pull'.

Rowing engages a large proportion of the body's muscles: the legs, back and arms all contribute to the effort required to propel the boat forward.

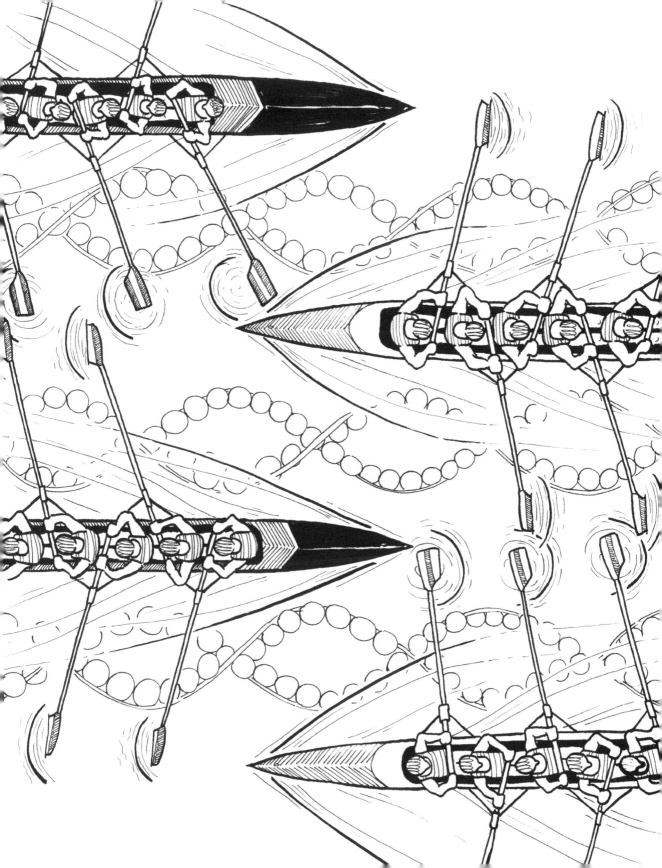

THE SKELETON

The skeleton is the scaffold of the body; it is what gives the body its shape. It can be divided into two divisions. The axial skeleton comprises the skull, vertebral column and ribcage. The appendicular skeleton is formed by the shoulder girdle, pelvic girdle and the limbs. It is attached to the axial skeleton.

In addition to supporting the body, the skeleton has many other functions. The joints, together with the muscles, allow for movement. The tough bone of the skeleton also protects many vital structures. The skull and facial bones protect the brain, the vertebral column protects the spinal cord, and the ribcage protects the heart and lungs, in addition to keeping the lungs from collapsing.

Avicenna (Ibn Sīnā in Arabic) was a physician-philosopher who lived in Persia during the eleventh century. He was one of the most influential figures of the Islamic Golden Age. One of his most important works was *The Canon of Medicine*, written in 1025. It remained one of the pre-eminent medical textbooks in Europe and the Islamic world for centuries. Illustrations of skeletons similar to the one here adorned his books.

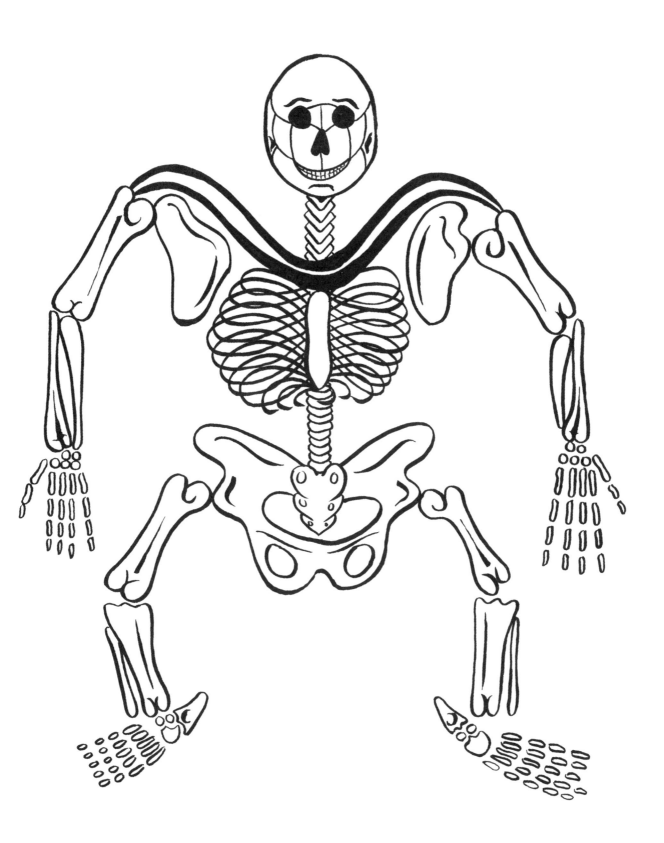

VERTEBRAL COLUMN

The spine, or vertebral column, is the central axis of the body. It carries much of our body weight, maintains our upright posture and protects the spinal cord. It is made up of thirty-three individual vertebrae. There are seven cervical vertebrae in the neck, twelve thoracic vertebrae and five lumbar vertebrae. The five sacral and four coccygeal vertebrae are fused.

The vertebral column is shaped with two curves. Lordosis is where the vertebral column curves forward in the cervical and lumbar areas (neck and lower back,

respectively). Kyphosis is where the column curves backwards, in the thoracic and sacral region (upper back and pelvis).

The spinal cord runs from the brain through the spinal canal. This canal is formed by the vertebral bodies in front, and a bony ring at the sides and behind. The bony exterior protects the cord from harm, while the gaps between individual vertebrae, known as foramina, allow the spinal nerves to exit the cord in order to travel to the rest of the body.

The Staff of Asclepius is a rod entwined by a single serpent, which was wielded by the Greek god of medicine and is a well-known medical symbol. It is often confused with the Caduceus, a staff entwined by two snakes and carried by Hermes, the messenger of the gods. This symbol mistakenly features in the logos of some healthcare bodies. This may originate from when it was used as a printers' symbol marked on books, including medical textbooks.

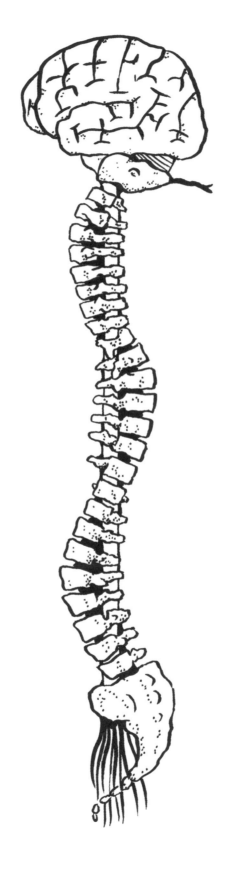

BONES OF THE SKULL

The skull is made up of two parts. The facial skeleton supports the face; the neuro-cranium houses and protects the brain. This part of the skull is made up of eight bones: the occipital bone, sphenoid bone, frontal bone and ethmoid bone, along with two parietal and two temporal bones. The bones are joined together by sutures.

In babies, the sutures are loose and allow the bones to slide over one another during birth. This is similar to the movement of the Earth's tectonic plates. However, as we grow up, the sutures become tougher, and by adulthood there is very little flexibility in the skull.

A condition called craniosynostosis is caused by the premature closure of these sutures. This causes deformity in the shape of the skull and a build-up of pressure on the brain. This condition often requires surgery.

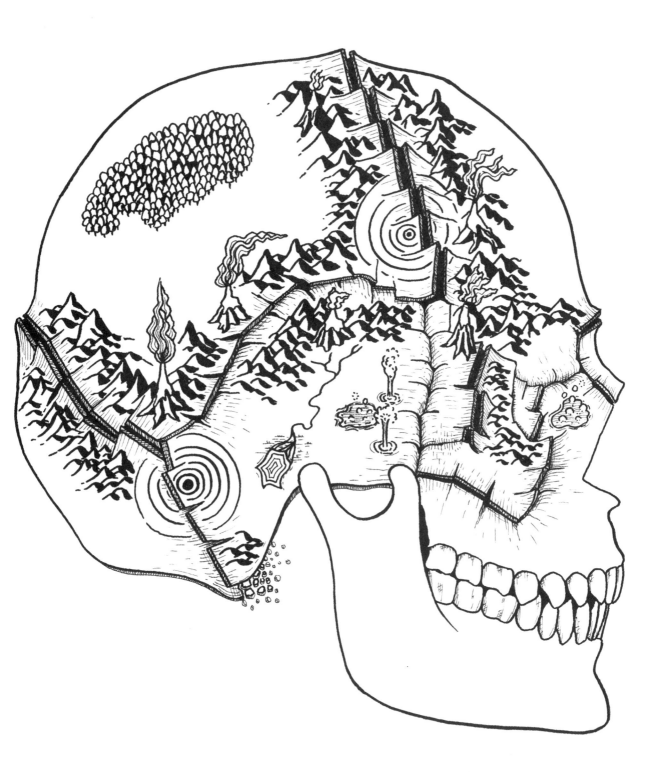

MUSCLES OF THE FACE

The face is made up of over twenty small muscles, supplied by a facial nerve on either side. These muscles are responsible for creating our facial expressions. This is everything from moving the eyebrows to puffing up the cheeks or pouting. If the facial nerve on one side is damaged, our ability to move these muscles is reduced and that side of the face will appear to droop. This can occur with a condition known as Bell's palsy.

People recovering from Bell's palsy can develop a condition called crocodile tears syndrome, the shedding of tears while eating. This is caused by abnormal recovery of the facial nerve, whereby the presence of food, instead of causing salivation, causes lacrimation (tearing up). It derives its name from the belief that crocodiles weep after killing their prey.

ATLAS AND AXIS

The skull is supported by two specially designed vertebrae, known as atlas and axis. Atlas, named after the Greek Titan, supports the skull and allows nodding movements. In other words, when you nod your head to say 'yes', that is thanks to atlas.

Atlas sits atop another specialised vertebrae called axis, the Latin for 'axle'. It acts as a pivot, allowing atlas to rotate from side to side. When you shake your head and say 'no', that is axis at work.

The junction between these vertebrae and the skull is one of the most critical body parts. The blood supply to your brain travels through these vertebrae, and the brainstem meets the spinal cord in them also. An injury at this point can leave a person without any function below the neck.

Atlas was a Greek Titan who was condemned by Zeus to hold up the celestial spheres in the West after the Titans lost their battle with the Olympian gods. He is commonly depicted holding a solid orb on his shoulders (as here), after a famous Roman statue known as the Farnese Atlas. This led to the common misconception that he held up the Earth itself, which was not the case.

PELVIC BONES

The pelvis is a basin-shaped ring of bone that connects the spine to the hips on either side. It is comparable to the shoulders of the arms. However, the shoulders are light and nimble to facilitate the wide range of movement of the arms. In contrast, the pelvis is more rigid, so as to withstand the forces of movement and to support the weight of the body while upright.

The pelvis is made up of three paired bones: the pubis, ilium and ischium. The pubic bones form the front of the pelvis.

The ilium is the largest part, and looks like a pair of wings protruding out of the pelvis. Several large muscles, including the gluteal muscles, attach to the ilium. The ischium is the bottom of the wings. The three bones meet to form the socket for the hip joint, known as the acetabulum.

These bones form a basin that contains the pelvic organs, i.e. the bladder, rectum, uterus, etc. In another similarity to a basin, the pelvis also has a hole in the bottom to let liquid out!

Males and females have a slightly different-shaped pelvis. The female pelvis is wider and shallower and is adapted for childbirth.

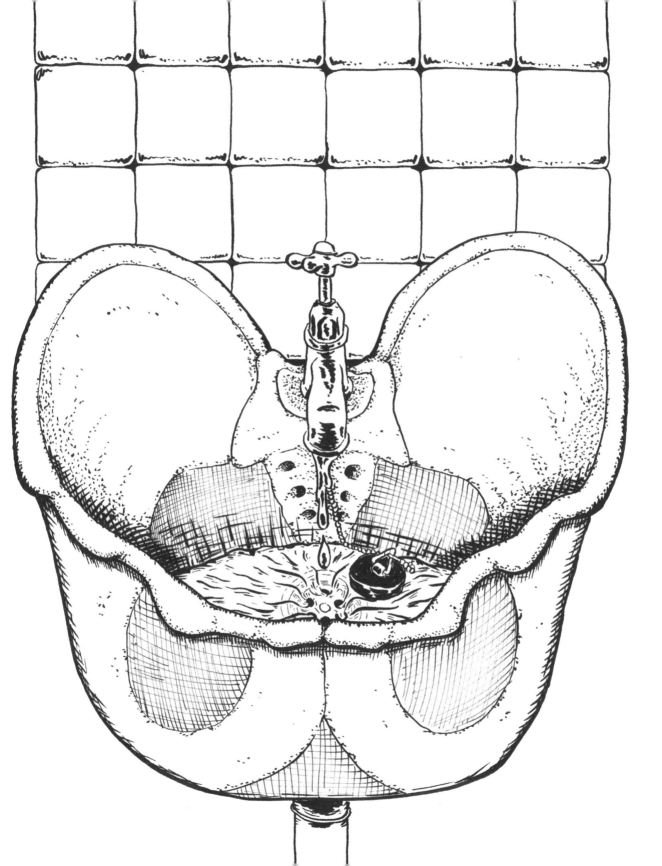

BONES OF THE UPPER LIMB

In anatomy, what most people call the arm is known as the upper limb. The arm is the part between the shoulder and elbow, with the forearm being the part between the elbow and wrist.

The shoulder girdle is made up of the clavicle (or collarbone) and scapula (or shoulder blade). The humerus is the major bone of the arm and forms a ball-and-socket joint at the shoulder.

The forearm is made up of two bones, the radius and ulna. These form a hinge joint with the humerus at the elbow.

The shoulder is a very mobile joint. It allows your arm a wide range of movement, from combing your hair to scratching your back. Part of the reason for this is that the socket aspect of the joint in the scapula is very shallow. A downside of this, however, is that shoulder dislocations are common. The weakest part of the joint is the front and as a result almost all dislocations are in this direction.

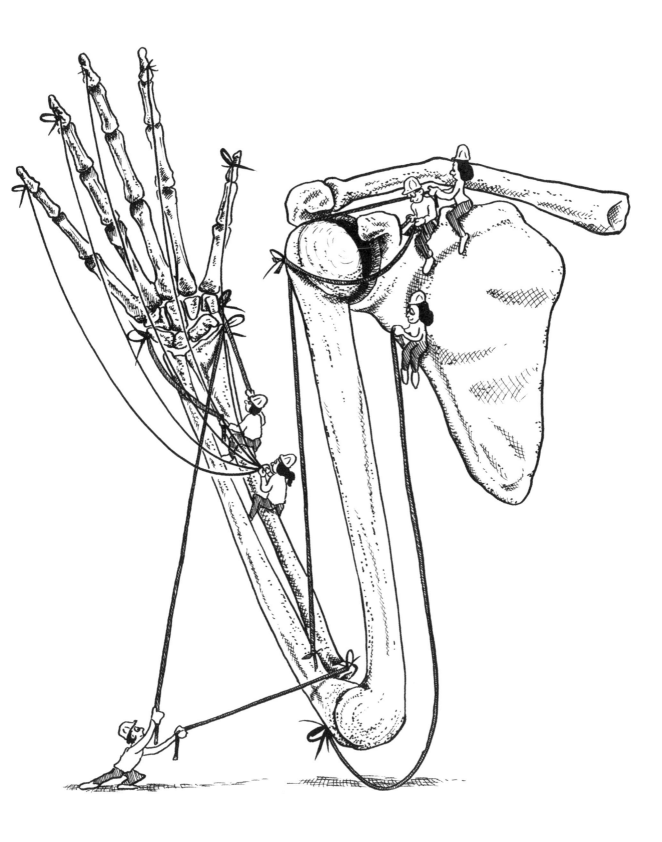

MUSCLES OF THE UPPER LIMB

The upper limb is adapted to balance strength and mobility. In comparison with the lower limb, it contains a vast number of muscles, allowing it greater range and dexterity of movement.

The shoulder has a number of muscles, including the pectoralis muscles in front, the trapezius and deltoid muscles above, and the latissimus dorsi muscle behind. The rotator cuff muscles lie underneath these muscles and control the rotation of the shoulder, as well as providing stability.

The elbow joint is moved by muscles in the arm. The biceps flex the elbow, while the triceps extend it. They are so-called because the bicep is made up of two muscle bodies, while the triceps is comprised of three.

The wrist and finger movements are mostly controlled by muscles in the forearm. The anterior forearm (the side of the palm of the hand) contains the muscles that flex – or bend – the fingers and wrist. They form tendons that travel into the hand and to the fingers. Overuse of these muscles can cause a condition called medial epicondylitis, or 'Golfer's Elbow', because they are the muscles used to make a golf swing. By contrast, the muscles on the posterior forearm (the side of the back of the hand) control extension of the wrist and fingers. Overuse of these muscles can cause lateral epicondylitis, or 'Tennis Elbow'.

During the Renaissance, artists such as Leonardo da Vinci and Andreas Vesalius created detailed studies based on their own dissections. These illustrations typically depicted figures with their skin, muscle layers and bones stripped off in layers. These figures came to be known as the écorché, French for 'flayed'.

THE HAND

Fitting with its status as one of the body's most important appendages, beneath the surface the hand is a complex structure. It is made up of twenty-seven bones. The eight small carpal bones form the carpal tunnel and wrist joint. The metacarpal bones articulate with the carpal bones and together form the palm of the hand. These articulate with the phalanges, which form the fingers and thumb.

Much hand movement is controlled by muscles in the forearm, which control the fingers through tendons. These tendons travel into the hand, attach to the bones in the fingers and control movement through a series of pulleys.

There are also a number of small muscles within the hand itself. The two most important groups are the thenar eminence by the thumb and the hypothenar eminence by the little finger. These allow these two digits a wider range of movement than the others. In between the digits are two groups of small muscles, the lumbricals and interossei muscles. These allow you to move your fingers in and out (as when you play scissors in rock-paper-scissors).

Finally, the skin of the hand is also special. The skin on the palm of the hand is tight and relatively thick, and can be bent along the creases. The ridges on the fingers, also known as fingerprints, increase friction and improve our grip. Our fingers also contain fat pads, which protect the important structures, nerves and tendons that would otherwise be vulnerable because of the lack of muscle.

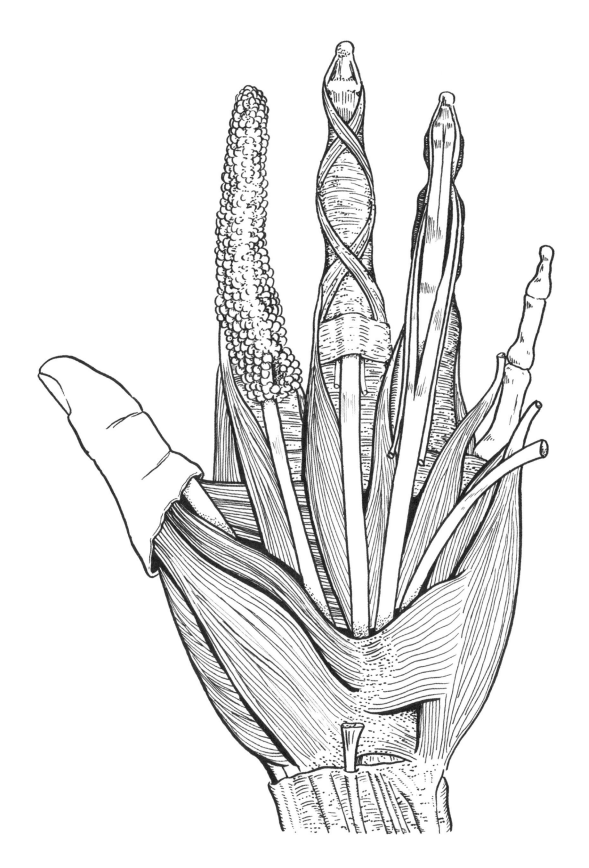

BONES OF THE LOWER LIMB

In anatomy, what is commonly called the leg is the lower limb. Between the hip and knee is the thigh, and between the knee and ankle is the leg. Beyond the ankle is the foot.

The femur forms the hip joint with the pelvis, a ball-and-socket joint similar to the shoulder. It is the longest bone in the body, and also one of the strongest. The leg is made up of the tibia and fibula, which articulate with the femur to form the knee joint. The patella, or kneecap, lies in the quadriceps tendon. It acts as a pulley, increasing the transmission of force from the quadriceps tendon.

The foot is similar to the hand. It is made up of small tarsal bones, which articulate with metatarsals and phalanges to form the toes. However, it is less flexible than the hand given its role in supporting body weight and movement.

The patella is a sesamoid bone, meaning it lies embedded within a tendon. There are other such bones in the body. For example, sesamoid bones are typically present within the tendons that flex the thumb and big toe, and can sometimes even be found in the joints of the smaller fingers and toes.

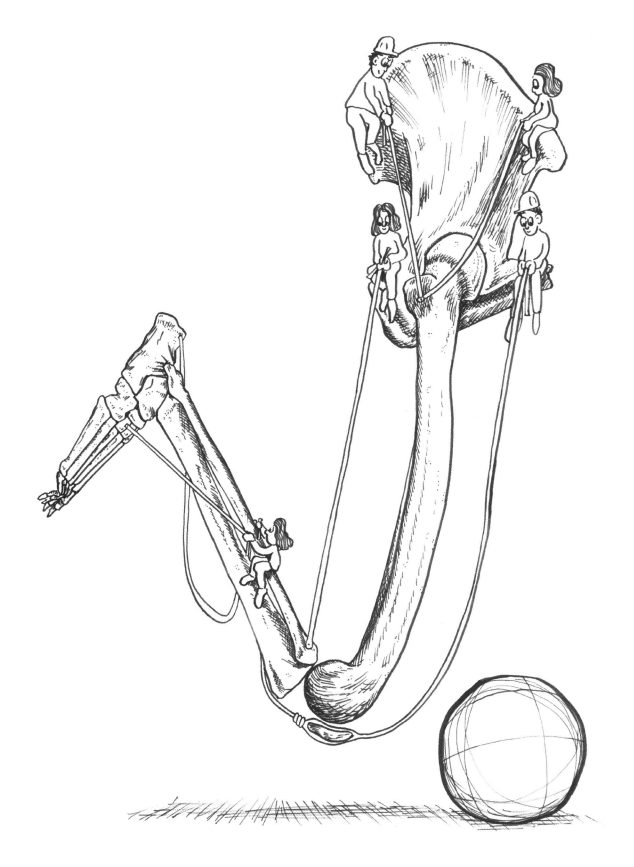

MUSCLES OF THE LOWER LIMB

The lower limb contains some of the largest and most powerful muscles in the body. They maintain our upright posture when standing or moving, and propel us forward.

The iliopsoas muscle extends from the pelvis to the hip joint and flexes the joint, such as when we do a high kick. The gluteal muscles are responsible for extending the hip joint, such as when we stand from a sitting position. The gluteus maximus and a smaller muscle called the tensor fascia lata contribute to the iliotibial band. This is a tough fibrous band that stretches from the iliac crest of the pelvis to the tibia on the outside of the lower limb. It is important for stabilising the knee.

There are three groups of muscles in the thigh. The quadriceps muscles are a group of four muscles (hence the name) that lie in the front of the thigh and form the patellar tendon. This tendon contains the patella, or kneecap, and inserts into the tibia. This allows us to extend the knee, such as when we kick a ball. The hamstrings lie in the back of the thigh and flex the knee and hip. The third group is the adductor muscles, which lie on the inside of the thigh. These muscles pull the leg inwards towards the midline.

The lower leg has two groups of muscles, which either move the foot towards or away from the ground (called plantar flexion and dorsiflexion respectively). The gastrocnemius and soleus muscles in the calf, for example, push the foot towards the ground, thereby propelling us forward.

ACKNOWLEDGEMENTS

This book would not have been created without hours and days of other people's time.

I must first thank the entire team at Mercier Press, who guided a novice like me through the process with patience and words of encouragement. Ever since I first sat down with Patrick, the commissioning editor, for a coffee in Clement & Pekoe to chat about some ideas, he has been a voice of encouragement helping me along. I must thank Alice, who is the brains behind the polished appearance of the book you hold. Thank you to Noel and Wendy, who stopped me from veering too far into medical-ese, and scrubbed the book of my many errors. Any that remain are entirely my own.

There are far too many colleagues and teachers from medical school and the hospitals around Ireland to thank individually. Saying that, I must thank David Cotter and Mary Cannon at the RCSI, who have been a source of mentorship and support for the past few years since we created *Journey Through the Brain*; and Clive Lee at the RCSI, who sat down with me to talk about anatomy and the history of art recently, but who also inspired my love of anatomy when I was a student in the RCSI.

Thanks to my many friends and family, who tolerated months of me hiding away over evenings and weekends, drawing at my desk. However, perhaps that was a blessing in disguise for them! To my parents, Mick and

Maura, who fostered a love of art and doodling in the margins and who knew something would come of it! I must also single out my friends Adam and Michael, in particular, who were a source for ideas and constructive criticism as the book took shape. This book would never have been completed without my partner, Dearbhla. Without her vast knowledge of medicine, creative thoughts and loving encouragement this book would remain unfinished.

ABOUT THE AUTHOR

Eoin Kelleher is a doctor, based in Dublin, training in anaesthesiology. He studied medicine in the Royal College of Surgeons in Ireland. When not putting people to sleep, he is at work. And when he is not at work, he is usually found at his drawing desk.

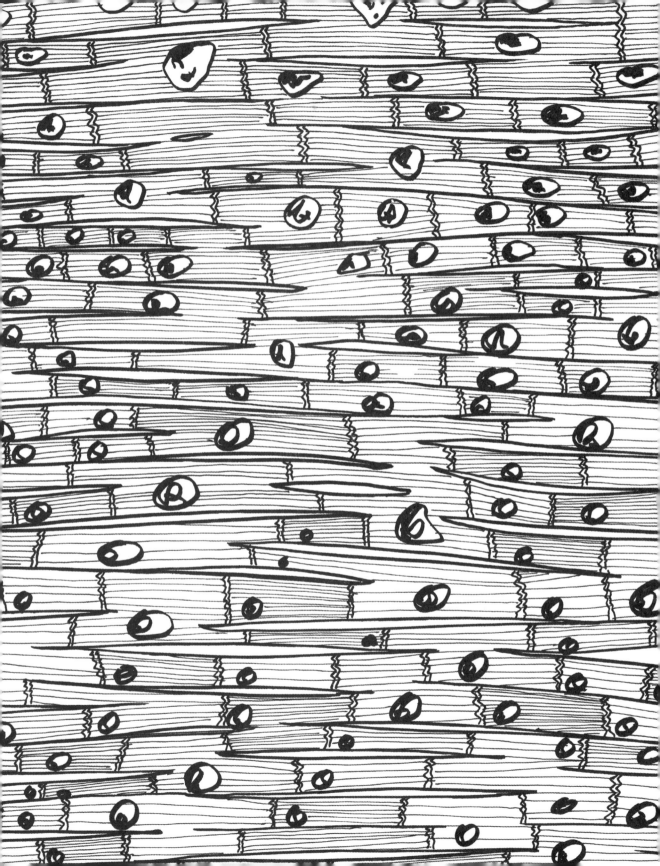